Understanding Greek and Roman Technology: From Catapult to the Pantheon

Stephen Ressler, Ph.D.

PUBLISHED BY:

THE GREAT COURSES
Corporate Headquarters
4840 Westfields Boulevard, Suite 500
Chantilly, Virginia 20151-2299
Phone: 1-800-832-2412
Fax: 703-378-3819
www.thegreatcourses.com

Printed in the United States of America

Stephen Ressler, Ph.D., P.E., Dist.M.ASCE

Professor Emeritus
United States Military Academy at West Point

Professor Stephen Ressler is Professor Emeritus at the United States Military Academy at West Point. He earned a Bachelor of Science degree from West Point, Master's and Ph.D. degrees in Civil Engineering from Lehigh University, and a Master of Strategic Studies degree from the U.S. Army War College. Professor Ressler is a registered professional engineer in the Commonwealth of Virginia. He is also an amateur artist and craftsman who created all of the physical models and most of the computer models used in this course.

Professor Ressler served for 34 years as a commissioned officer in the U.S. Army Corps of Engineers and retired at the rank of brigadier general in 2013. He served in a variety of military engineering assignments in the United States, Europe, and Central Asia, including 21 years as a member of the West Point faculty. At West Point, he taught courses in engineering mechanics, structural analysis, structural design, construction management, and civil engineering professional practice and history. In 2007, he deployed to Afghanistan to develop a civil engineering program for the newly created National Military Academy of Afghanistan in Kabul. In that capacity, he designed the civil engineering curriculum, hired the first cohort of Afghan faculty, and developed two laboratory facilities.

Professor Ressler has focused his scholarly work and professional service in the area of engineering education. He has written more than 80 scholarly papers on teaching techniques, faculty development, curriculum assessment, engineering outreach to primary and secondary schools, engineering accreditation, and information technology. His work has earned nine Best Paper Awards from the American Society for Engineering Education.

Professor Ressler is the creator and director of the West Point Bridge Design Contest (see http://bridgecontest.usma.edu), a nationwide Internet-based competition that has introduced engineering to more than 50,000 middle school and high school students since 2001. He is also a developer and principal instructor for the Excellence in Civil Engineering Education (ExCEEd) Teaching Workshop, a landmark faculty development program sponsored by the American Society of Civil Engineers (ASCE). The workshop has provided rigorous teacher training to more than 500 civil engineering faculty members from more than 200 colleges and universities over the past 14 years.

Professor Ressler has won numerous prestigious national awards. From ASCE, he received the President's Medal and the ExCEEd Leadership Award. ASCE also named him a Distinguished Member in 2005. From the American Society for Engineering Education, he received the George K. Wadlin Distinguished Service Award, the Distinguished Educator Award, and the Dow Outstanding New Faculty Award. He also received the Society of American Military Engineers' Bliss Medal for Outstanding Contributions to Engineering Education, the American Association of Engineering Societies' Norm Augustine Award for Outstanding Achievement in Engineering Communications, the Premier Award for Excellence in Engineering Education Courseware, and the EDUCOM Medal for application of information technology in education.

Professor Ressler was one of *Engineering News-Record*'s Top 25 Newsmakers Who Served Construction in 2000. Most recently, he received ASCE's highest award—the Outstanding Projects and Leaders (OPAL) Award for 2011. The OPAL Award is presented to only five of ASCE's 140,000 members each year. ■

Table of Contents

INTRODUCTION

LECTURE GUIDES

Table of Contents

Table of Contents

SUPPLEMENTAL MATERIAL

Disclaimer

This work is not the product of an official of the United States Army acting in his official capacity. The contents of this work are not to be considered as the official views of the United States Military Academy, United States Army, or Department of Defense. Neither this product nor its content are endorsed by the United States Military Academy, United States Army, or Department of Defense.

Understanding Greek and Roman Technology: From Catapult to the Pantheon

Scope:

When you consider the great civilizations of classical antiquity, what's the first thought that comes to mind? If you're culturally minded, perhaps it's a particular piece of Greek sculpture, a tragedy by Sophocles, one of Plato's dialogues, or Homer's *Iliad* or Virgil's *Aeneid*. If you're into politics, maybe Athenian democracy or the Roman Republic comes to mind. If you like sports, you may be thinking of the Olympic Games or bloody gladiatorial contests in the Roman arena. If you're a military history buff, you're probably imagining a great battle, such as Marathon, Thermopylae, or Actium.

But there is another aspect of the ancient Greek and Roman world that also deserves your attention: technology. The overarching goal of this course is to augment and enrich your appreciation for the cultural, political, and historical dimensions of classical antiquity by providing you with an opportunity to explore its technological dimension from an engineering perspective.

Greek and Roman technology is worth learning about for three reasons:

- First, technology can sometimes influence the course of human events quite directly. But even when it doesn't, technology always reflects the social, political, and cultural context from which it emerged. Thus, we can appreciate an ancient civilization more deeply if we understand something about the technological developments it fostered.

- Second, we should know about Classical-era technology because it has influenced our modern world in many substantive ways. We will encounter many modern legacies of ancient technology throughout this course.

- Third, ancient technology is incredibly interesting and sometimes even astonishing! This is particularly true because the design of ancient technological systems was constrained by relatively crude materials, extremely limited sources of power, and a lack of scientific models that could be used to predict the behavior of physical systems (as scientific models are used today). As a result, ancient engineers had to be exceptionally clever in designing structures and mechanical devices, and these systems often display the ingenuity of their creators with great clarity. Learning to discern and appreciate this virtuosity in design is a great source of joy in studying ancient technology.

The scope of this course is limited primarily to the development of large-scale engineered systems during the period of classical antiquity, though we will occasionally look at earlier technologies when they are relevant to the subject at hand.

The term "classical antiquity" refers to Greco-Roman civilization during a 1,300-year period that spans five major historical eras: the Greek Archaic era, which began around 800 B.C., when Homer's epic poems were first written down; the Hellenic era, which began with the emergence of Athenian democracy around 500 B.C. and ended with the death of Alexander the Great in 323 B.C.; the Hellenistic era, characterized by the spread of Greek influence through much of the Mediterranean world under the auspices of competing kingdoms established by Alexander's successors; the Roman Republic, which was established in 509 B.C. and initially coexisted with the Hellenistic kingdoms but ultimately conquered them all; and finally, the Roman Empire, which is generally dated from 27 B.C., when Octavian assumed the title Augustus, to A.D. 476, when the last western Roman emperor was deposed.

Although our course spans this 1,300-year period, its organization is thematic rather than chronological. These lectures are organized into three major sections, each addressing a particular category of engineered system. First, we'll spend eight lectures examining important structural and construction technologies in buildings from the earliest Greek temples to the Roman Pantheon of the 2nd century A.D. We'll then devote six lectures to Greek and Roman infrastructure systems—roads, bridges, water supply systems, sewage systems, and public baths—as well as the urban planning methods that tied all these systems together. Finally, we'll have eight lectures on ancient machines used in construction, water-lifting, power production, grain-milling, warfare, and transportation.

This is not a survey course. For this reason, we will not attempt to cover every possible category of ancient technology. Such topics as mining, agriculture, timekeeping, tools, military arms, coinage, textiles, ceramics, and glass are both interesting and important, but they are beyond the scope of this course on large-scale engineered systems.

This is an engineering course. Thus, in each technological category, we will examine just a few representative examples in depth. For example, hundreds of Greek temples have survived from antiquity, yet when we consider this topic in Lecture 4, we'll spend most of our time exploring just one building—the Temple of Concordia at Agrigento, Sicily. This is neither the largest nor the most important Doric temple ever built, but it is one of the best preserved and best documented. Thus, we can use the Temple of Concordia as a vehicle for learning deeply about the Greek temple as a technological entity, rather than surveying many examples more superficially.

For all of the engineered systems we examine, we will seek to answer three "big questions": How was it built? How did it work as an engineered system? And how is it situated within the broader context of technical development in the ancient world?

To answer these questions rigorously, we will need to familiarize ourselves with some basic scientific and engineering principles throughout the course. In our lectures on structures, we'll learn some basic concepts in engineering mechanics—how structural elements, such as beams, columns, and arches, carry load. When we examine infrastructure systems, we'll learn about hydrology, hydraulics, and surveying. In our study of ancient machines, we'll apply the concepts of energy, work, power, mechanical advantage, and buoyancy. Even if math and science aren't your strong suits, however, you'll find that the realm of ancient technology provides a compelling and rewarding context for learning some of these basic concepts about how our world works.

We will conclude the course with a lecture devoted to the modern legacy of ancient technology. Through this exploration, you will acquire tools and perspectives that will allow you to continue appreciating Greek and Roman technology even after our course has ended. ■

Technology in the Classical World

Lecture 1

A study of the great civilizations of classical antiquity presents many fascinating areas of focus—culture, politics, sports, military history. The goal of this course, however, is to explore technology in the classical world. A key figure in our journey is Marcus Vitruvius Pollio, the author of *De Architectura*—the sole surviving treatise on architecture and engineering from the ancient world. Vitruvius was a 1st-century-B.C. Roman *architectus*—a job title that encompassed not only architecture but also the modern professions of engineer, construction manager, and urban planner. Vitruvius suggests that the work of the *architectus* requires a learned, broadly educated person; in the spirit of The Great Courses, studying the work of the *architectus* will help us expand our knowledge, as well.

The Battle of Salamis

- One of the most pivotal events in human history was the Battle of Salamis, which occurred in September 480 B.C. King Xerxes of Persia had invaded Greece with an immense multinational army and a supporting fleet of more than 1,000 warships. On land, Xerxes' army crushed the Greeks' forward defense at Thermopylae pass and then marched south to sack Athens and seize all of Attica except Salamis Island.

- But at Salamis, a small Greek fleet lured the mighty Persian armada into the narrow strait separating the island from the mainland and won a stunning victory, despite being outnumbered three to one. The Battle of Salamis cost the Persians some 300 ships and tens of thousands of crewmen. It dealt a deathblow to the Persian invasion—and opened the door to the extraordinary cultural achievements of Classical Greece, changing the course of world history.

- The warships on both sides of the battle were **triremes**—125-foot wooden galleys, each rowed by 170 men arrayed on three levels. The Greeks won at Salamis for a number of reasons.

- They employed a superior strategy—choosing to fight in the narrow channel, which negated the Persians' advantage in numbers.

- The Greek cultural predisposition toward *metis*, or cunning, was also a factor, as the Athenian general Themistocles used a brilliant ruse to lure the Persians into the channel.

The trireme was, in essence, a human-powered torpedo; it engaged enemy vessels by attacking them with the bronze-clad ram on its prow.

- Politics influenced the outcome, as well. Most of the Persian fleet was manned by conquered peoples. The historian Herodotus tells us that the captains of the Persian fleet repeatedly made poor tactical decisions because they were preoccupied with impressing the king. And their oarsmen—subjects of a despotic ruler—could hardly have rowed with the same spirit as the Greeks, who were free citizens whose families and cherished homeland hung in the balance.

The Greeks' Superior Technology at Salamis

- A full explanation of the Greeks' victory at Salamis must take into account how it was influenced by **technology**. Even though both sides used the same type of warship—the trireme—the Greeks and Persians employed this technological system quite differently.

- The night before the battle, the Greeks pulled their triremes onto shore to dry out their wooden hulls and protect them from worms.

The Persian ships remained at sea—a move that ignored two key technological limitations of the trireme. First, without having been dried out on shore, the Persian hulls would have been waterlogged and, therefore, slower than their Greek adversaries'. And, more important, the Persian oarsmen—the human engines of the trireme—would have rowed into battle without having slept for a full 24 hours.

- For the sake of optimized performance, the trireme was built just barely large enough to accommodate its 170 rowers seated on wooden benches. While a healthy adult can produce a sustained power output of about 1/10 horsepower, a well-rested man in good physical condition can produce more than 1 horsepower for short periods. Given this tenfold range in potential human power output, the fatigue of the Persian crews would have resulted in a substantial degradation of their ships' performance.

- Furthermore, Herodotus tells us that Xerxes augmented each trireme's standard crew with 30 Persian soldiers, additional troops that would have increased the overall weight of each ship by about 8 percent. These men, stationed on the ship's upper deck, would also have made the vessel dangerously top-heavy—a situation exacerbated by the fact that triremes carried no ballast.

- There can be no doubt, then, that the Greek victory against such incredible odds can be attributed, at least in part, to the combined effects of the Persian ships' overloaded, unstable, waterlogged hulls and the reduced power output of tired oarsmen. These factors are all associated, in some way, with the unique technological characteristics of the trireme.

Why Study Ancient Technology?

- There are three main reasons to study ancient technology. First, as we saw at Salamis, technology can sometimes directly influence the course of human events. But even when it does not, technology always reflects its social, political, and cultural context. Thus, we can appreciate an ancient civilization more profoundly if we understand the technological developments it fostered.

- Second, Greek and Roman technologies have influenced our modern world in many substantive ways.

- Third, ancient technology is, at times, truly astonishing—given the fact that the design of ancient technologies was severely constrained by relatively crude materials, extremely limited sources of power, and a lack of scientific models to predict the behavior of physical systems. As a result, ancient engineers had to be exceptionally clever in designing **structures** and mechanical devices; these systems often display the ingenuity of their creators with great clarity.

Scope and Organization of the Course

- The scope of this course is limited primarily to the development of large-scale engineered systems during the period of classical antiquity, which refers to Greco-Roman civilization during a 1,300-year period that spans five major historical eras:
  - **Greek Archaic period**, which began around 800 B.C., when Homer's epic poems were first composed.

  - **Hellenic period** (also called Classical Greece), which began with the emergence of Athenian democracy around 500 B.C. and ended with the death of Alexander the Great in 323 B.C.

  - **Hellenistic period**, characterized by the spread of Greek influence through much of the Mediterranean world under the auspices of competing kingdoms established by Alexander's successors. It began with the death of Alexander the Great in 323 B.C. and ended with the Roman conquest of Egypt in 30 B.C.

  - **Roman Republic**, which was established in 509 B.C. and initially coexisted with the Hellenistic kingdoms but ultimately conquered them all.

- Roman Empire, or imperial Rome, which is generally dated from 27 B.C., when Octavian assumed the title Augustus, to A.D. 476, when the last western Roman emperor was deposed.

- These lectures are organized into three major sections, each addressing a particular category of engineered system:
  - Important structural and construction technologies in buildings from the earliest Greek temples to the Roman Pantheon of the 2nd century A.D.

  - Greek and Roman **infrastructure** systems—roads, bridges, water supply, sewage systems, and public baths—as well as the urban planning methods that tied all these systems together.

  - Ancient **machines** used in construction, water transport, power production, the milling of grain, warfare, and transportation.

- Because this is an **engineering** course, in each technological category, we'll examine a few representative samples in depth. For all the engineered systems we examine, we'll seek to answer three questions: How was it built? How did it work as an engineered system? How is it situated within the broader context of technical development in the ancient world?

Four Phases of Technological Development

- There was a persistent belief throughout much of the 20th century that the Classical era was a time of technological stagnation. This theory was based primarily on the observation that the Greco-Roman world produced few fundamentally new inventions.

- Indeed, this premise is quite correct. Inventions that predated the Greek Archaic period include animal power, metallurgy, coinage, stone masonry, terra-cotta, wheel, wedge, lever, pulley, sail, and even the arch (a technology we tend to associate with ancient Rome that is actually much older). But during the entire 1,300-year span of classical antiquity, the only major new inventions were the screw, water wheel, and concrete.

- Does lack of invention really constitute technological stagnation? Many current Classical scholars say no. They argue that proponents of the technological stagnation theory have defined "technological development" far too narrowly—as "invention" and nothing more. A recent, more holistic model defines technological development in terms of four phases:
    - **Invention**—the act of implementing an original idea in a new device.
    - **Innovation**—the process by which an invention is brought into use.
    - **Diffusion**—the process by which an innovation is communicated through a social system.
    - **Technology in use**—the processes of employing existing technologies and adapting them to new purposes over time.

- In terms of this four-phase model, it is evident that the Greeks and the Romans did little inventing, but they did contribute immeasurably to technological development through innovation, diffusion, and use. For example, the Greeks did not invent the lever, but their integration of the lever into the design of the bronze force pump was a brilliant innovation. Subsequent diffusion of this device throughout the Mediterranean world resulted in its adaptation to an incredible variety of uses.

- The ancient engineers were astonishingly ingenious, and their work reflects great creativity in design, deep qualitative understanding of engineering principles, and a well-honed ability to translate ideas into functioning products.

Museum of Alexandria

- The most productive periods of Classical-era technological development occurred under the patronage of powerful political leaders. In the modern world, we have a system of patent law, enabling inventors to reap direct economic benefits from their work.

But in a world without patents, royal patronage was practically the only way to provide a similar incentive. Thus, as we'll see, technological development was far more robust in the kingdoms of the Hellenistic world than it had been in the democratic city-states of Hellenic Greece, and the engineering achievements of imperial Rome far exceeded those of the Roman Republic.

- Political patronage of science and technology is best exemplified by the famed **Museum of Alexandria**, established by the Ptolemaic kings of Egypt around 300 B.C. No physical remains of the museum have survived, but we know from ancient texts that this "House of the Muses" served as a research center that brought together some of the Hellenistic world's finest scholars to solve practical problems. During the coming lectures, we'll meet several of these men—Ctesibius, Philo of Byzantium, Hero of Alexandria—and we'll examine the important technologies they developed at this extraordinary institution.

- In the next lecture, our journey begins with an introduction to engineering materials—the substance of technology.

Important Terms

diffusion: The process by which an innovation is communicated through a social system.

engineering: The application of math, science, and technology to create a structure, device, machine, system, or process that meets a human need.

Greek Archaic period: Historical period that began around 800 B.C. and ended with the emergence of Athenian democracy around 500 B.C.

Hellenic period (also called Classical Greece): Historical period that began with the emergence of Athenian democracy around 500 B.C. and ended with the death of Alexander the Great in 323 B.C.

Hellenistic period: Historical period that began with the death of Alexander the Great in 323 B.C. and ended with the Roman conquest of Egypt in 30 B.C.

infrastructure: Large-scale technological systems that enhance societal functions, facilitate economic development, and enhance quality of life.

innovation: The process by which an invention is brought into use.

invention: The act of implementing an original idea in a new device.

machine: An assembly of fixed or moving parts, used to perform work.

Museum of Alexandria: A Hellenistic institution of learning, research, and invention, established by the Ptolemaic kings of Egypt around 300 B.C. The word *museum* refers to a "house of the Muses," rather than a museum in the modern sense.

Roman Empire, imperial Rome: Historical period that is generally considered to have begun with Octavian's assumption of the title *Augustus* in 27 B.C. and ended in A.D. 476, when the last western Roman emperor was deposed.

Roman Republic: Historical period that began with the establishment of the Roman Republic in 509 B.C. and is generally considered to have ended with Octavian's assumption of the title *Augustus* in 27 B.C.

structure: A technological system—typically a building, bridge, or tower—that is designed to carry load.

technology: A human-made structure, device, machine, system, or process that meets a human need. Technology is the product of engineering.

technology in use: The processes of employing and maintaining existing technologies, and adapting them to new purposes over time.

trireme: In antiquity, a warship with oars on three levels. In later eras, other trireme configurations were also used, for example, one level of oars with three rowers per oar.

Suggested Reading

Oleson, *The Oxford Handbook of Engineering and Technology*, chapter 3.

Strauss, *Salamis*.

Vitruvius, *The Ten Books on Architecture*, chapter 1.

Questions to Consider

1. Why is it overly restrictive to define "technological development" solely in terms of invention?

2. What modern events have been strongly influenced by technology?

3. How does modern technological development reflect the economic, social, and political character of our modern world?

The Substance of Technology—Materials

Lecture 2

In Deuteronomy 8, Moses tells the Hebrews: "For the Lord your God is bringing you into a good country, a land with springs and fountains welling up in the hills and valleys, a land whose stones contain iron, and in whose hills you can mine copper." Materials mattered in the ancient world. Entire civilizations rose and fell on the basis of their ability to exploit materials—to enhance their military, economic, and political power. It is hardly surprising, then, that the ages of human history are customarily defined in terms of their dominant materials—Stone Age, Bronze Age, Iron Age. In that sense, the story of human civilization is the story of materials.

Mechanical Properties of Materials

- **Mechanical properties** are characteristics that describe how a material responds to **forces**. A force is simply a push or pull, defined in terms of both magnitude and direction.

- Think of a simple stone **column**—a structural element we might find in a Greek temple or Roman basilica. When we apply a compressive force to the column, it shortens. At the microscopic level, the individual atoms in its crystal structure are pushed closer together. But the bonds between these atoms are quite strong, and they resist being deformed. The effect of these interatomic bonds resisting deformation is the phenomenon we call **stress**.

- Mathematically, stress is expressed as force per area (for example, pounds per square inch). The area in question is the **cross-sectional** area. Stress comes in two "flavors": **tension** (elongation) and **compression** (shortening). The tensile and compressive strengths of a given material are not necessarily the same; for example, stone is much stronger in compression than in tension.

- The concept of stress is extremely important, because the most important mechanical property of a material—its **strength**—is defined in terms of stress. Specifically, the strength of a material is defined as the maximum stress the material can withstand before it breaks.

Stone and Wood: The Earliest Materials

- When our distant ancestors first began manipulating materials for practical purposes, they naturally used what they found lying about: stone and wood. Indeed, the oldest known human-made artifacts were simple stone tools—really just sharp chips of rock—used for cutting or chopping. Examples dating from 2 million years ago have been found in east Africa.

- By the time our own species, *Homo sapiens*, emerged roughly 200,000 years ago, humans were making hand axes consisting of a stone blade bound to a wooden handle with strips of hide or cemented with tree resin. This is a very early example of combining two materials, each used in a manner consistent with its mechanical properties.

- Through the ages, the use of stone and wood advanced considerably. The extent of this advance is evident in the Parthenon of Classical-era Athens, with its elegantly proportioned marble columns and elaborate wooden roof structure. Yet hundreds of millennia after the invention of the primitive hand ax, the Parthenon still reflects the same inherent strengths and limitations of stone and timber.

- Because stone is strong in compression, it is well suited for columns and walls, but it is weak in tension, so it is not well suited for **beams**—which explains why the columns of the Parthenon are spaced so closely together. Stone is also quite workable; the marble of the Parthenon could be carved with great precision. And stone is durable, allowing us to enjoy this architectural marvel 2,400 years after it was built.

- In contrast, timber is reasonably strong in both tension and compression; thus, it is much better suited for structural components subject to bending, such as roof beams. Wood is easily cut and joined with simple tools; however, wood lacks **durability** and is highly vulnerable to rot and fire. For this reason, no Greek temple roofs have survived to modern times; we can only speculate about their configuration based on the stone sockets in which the wooden beams were supported.

Clay: The First Manufactured Material

- In addition to stone and wood, a third material readily available to prehistoric humans was clay. The distinguishing property of clay is its **workability**—the ease with which it can be molded into any shape. This property reflects clay's unique atomic structure, consisting of thin layers of atoms that are only weakly bonded to each other.

- Because of clay's plasticity, early builders learned that they could mold it into mud bricks, made by mixing clay with sand and straw, then packing the mixture into wooden molds and placing them in the sun to dry. Though mud bricks are easy to produce, they make a poor structural material: Sun-dried clay has only low to moderate compressive strength, essentially zero tensile strength, and poor durability.

- Of course, the poor durability of sunbaked clay can be overcome by firing it in a kiln. When clay is subjected to intense heat, its original layered atomic structure changes to a more rigid three-dimensional crystal structure. The result is a ceramic—a crystalline material that is strong, heat resistant, and water resistant. Clay was the first material that humans were able to transform into a fundamentally new substance; thus, fired clay, or **terra-cotta**, was, in effect, the first true manufactured material.

Copper: The First Manufactured Metal

- The first manufactured metal was copper. We have evidence of its use as early as 9000 B.C. in the Middle East, although these artifacts were apparently made from lumps of relatively pure copper found in streambeds. It was another 5,000 years before the production of copper from mineral ores began. But these ores could not be exploited until ancient metallurgists developed the process we call **smelting**—the heating of copper ore with **charcoal** to produce pure copper. Smelting occurs through a chemical reaction called **oxidation-reduction**.

- Copper smelts at 2,000° F, which is considerably lower than the temperature required to melt iron (about 2,800° F). This explains why copper was the first metal to be exploited by humans, even though iron is about 1,000 times more plentiful in the earth's crust.

- Metals are relatively strong, and they have equal strength in tension and compression. Metals can also be formed by **casting**—by heating to the melting point, then pouring the molten material into a mold. But most important, metals have a high degree of **malleability** and are actually strengthened by hammering—a characteristic that makes them particularly well suited for tools and weapons.

- All these desirable characteristics come at a price: Smelting copper consumes a vast amount of charcoal fuel. In antiquity, **colliers** used about 7 pounds of wood to make 1 pound of charcoal, and smelters used 20 pounds of charcoal to smelt 1 pound of copper. The grand total is 140 pounds of wood to smelt a single pound of copper. As metals became ever more popular, the resulting insatiable demand for charcoal caused widespread deforestation. The barren terrain of modern Greece was caused largely by the clear-cutting of forests during the Classical era and subsequent erosion of irreplaceable topsoil.

Bronze Supersedes Copper

- Despite the depletion of timber, experimentation with metals continued. Around 3200 B.C., smelters began adding small amounts of tin to copper to produce a new material: **bronze**.

- A metal composed of two or more elements is called an **alloy**; bronze is an alloy of copper and tin. Interestingly, adding an impurity to a substance often increases its strength; thus, an alloy is typically stronger than its constituent elements. Further, because the melting temperature of bronze is 140° F lower than that of pure copper, it is actually better suited for casting.

- Bronze quickly superseded copper as the preferred metal of ancient civilizations. It held that distinction for more than a millennium and, in the process, gave its name to a new era: the Bronze Age. During this period, Homeric heroes were clad in bronze, while bronze plows, sickles, and saws improved the lives of common people. By the 10th century B.C., bronze-casting technology was sufficiently advanced that an artisan named Hiram of Tyre was able to fabricate 27-foot-tall bronze pillars for Solomon's Temple in Jerusalem.

The Iron Age

- By that time, however, bronze was already giving way to a new material—iron—that would define another new age. The earliest iron known to humans actually originated in meteorites and was used as a "luxury metal" for several thousand years before the Iron Age began. But sometime in the 2nd millennium B.C., smelters in the Near East developed techniques for producing iron from an ore called **hematite**.

- Although the chemical reaction theoretically produced pure iron, the actual product of these crude ancient furnaces was far from pure. Iron melts at about 2,800° F, but ancient forges could not do much better than 2,200° F. Thus, the smelting process produced a spongy mass composed of iron,

Iron was used for clamps in stone masonry construction; spikes for fastening ships' hulls together; and axles, fittings, and connectors in such machines as the waterwheel.

unreacted ore, unburned charcoal, and other impurities. While the mass was still red-hot, it had to be hammered vigorously to create a usable material. This product was called **wrought iron**, because it had been wrought (or worked) through this hammering process.

- In antiquity, iron was not a significant improvement over bronze, yet the Bronze Age did yield to the Iron Age. The reason was, quite simply, availability. Like most new technologies, iron found its earliest applications in warfare—primarily for weapons and armor. But as its production cost decreased, iron was adopted in all sorts of civil technologies. In a very real sense, the Classical world was held together with iron.

The Genius of Ancient Engineering

- All the materials discussed in this lecture were well established before the Classical era. Two more materials, concrete and lead, came into widespread use during this era. However, even when we add concrete and lead, the materials available to Greek and Roman builders were still limited, both in number and in their suitability for engineering applications—in sharp contrast with the modern world's immense and ever-expanding array of custom-manufactured metals, plastics, and ceramics. We credit the genius of ancient engineers in their ability to use extremely limited materials in ways that capitalized on their unique strengths and compensated for their inherent weaknesses.

- Although we tend to think of materials as enablers of technological development, we have also seen how ancient systems for acquiring and processing materials are important examples of technological development in their own right. We'll explore this idea in much greater depth in the next lecture, when we examine how that most ubiquitous of all ancient construction materials, stone, was quarried, transported, shaped, and assembled into masonry—the very fabric of the ancient world's most beautiful and enduring structures.

Important Terms

alloy: A metal composed of two or more elements.

beam: A structural element that carries load primarily in bending.

bronze: An alloy of copper and tin.

casting: The process of forming a metal into a desired shape by heating it to the melting point and then pouring the molten material into a mold.

charcoal: A fuel produced by heating hardwood in a reduced-oxygen environment to drive off water and resins. Because charcoal is nearly pure carbon, it is capable of burning at much higher temperatures than wood.

collier: A person who produces charcoal.

column: A structural element that carries load primarily in compression.

compression: An internal force or stress that causes shortening of a structural element.

cross-section: The geometric shape of a structural element, viewed from its end.

durability: The capacity of a material to resist deterioration by weathering, corrosion, or rot.

force: A push or pull applied to an object. A force is defined in terms of both magnitude and direction.

hematite: A type of iron ore.

malleability: The extent to which a material can be shaped or formed into thin sheets by hammering. Metals are generally very malleable, in contrast with stone, clay, and concrete.

mechanical properties: Characteristics of a material that describe how the material responds to forces.

oxidation-reduction: A chemical reaction in which one substance loses electrons (and is said to be oxidized) and another gains electrons (and is said to be reduced).

smelting: The process of heating an ore to produce a usable metal.

strength: The maximum stress a material can withstand before it breaks. Strength can be defined for both tension and compression.

stress: The intensity of internal force within a structural element, defined in terms of force per area (pounds per square inch).

tension: An internal force or stress that causes elongation of a structural element.

terra-cotta: A ceramic material created by firing clay in a kiln to improve its strength and durability.

workability: The extent to which a material can be molded, carved, or deformed to attain a desired physical shape.

wrought iron: Iron produced by hammering the product of the smelting process (called a bloom) to drive out impurities.

Suggested Reading

Riley, Sturges, and Morris, *Statics and Mechanics of Materials*, chapters 2–4.

Salvadori, *Why Buildings Stand Up*, chapters 4–5.

Sass, *The Substance of Civilization*, chapters 1–5.

Questions to Consider

1. Why are the first three ages of human history—Stone Age, Bronze Age, and Iron Age—named for materials?

2. Why is modern materials science considered to have originated with ancient terra-cotta?

3. Under what circumstances might the selection of a material for an engineered system not be based on the material's strength?

4. Why did the malleability of metal stimulate the widespread development of processes for smelting copper, bronze, and iron?

From Quarry to Temple—Building in Stone

Lecture 3

When we look at such architectural gems as the Temple of Athena Nike in Athens, we tend to focus on the building's beautiful proportions, its elegant fluted columns, its finely crafted Ionic capitals, and the delicate carvings of its entablature. But the temple's plain masonry walls are equally intriguing when we consider their construction: how the blocks were extracted from solid bedrock at a time when the most sophisticated quarrying tool was a simple iron-tipped pick; how the blocks were squared, shaped, and fitted so perfectly together with only hand tools and without mortar; and how the blocks were lifted into place and positioned with such precision.

The Greeks Perfect Ashlar

- Our story begins in the 2nd millennium B.C., during the ascendency of the great Bronze Age civilization of Mycenae in the northeastern Peloponnese. The Mycenaeans developed a form of monumental stone masonry, created by piecing together large polygonal blocks of stone in irregular patterns, with minimal cutting and fitting.
  - Later Greeks called this type of construction "**cyclopean**," because they believed that only the mythical Cyclops could have been strong enough to move such enormous stones.

  - With the collapse of Bronze Age civilization around 1200 B.C., however, this technological know-how was lost—and not rediscovered until centuries later.

- Around the late 8th century B.C., during the Archaic period (the same time that the polis emerged as a political entity), Greeks once again began constructing public buildings of stone. This development was probably stimulated by a desire for buildings with greater dignity and permanence, but the move may have been influenced by deforestation—the shortage of timber resulting from increasing demands in metal production and shipbuilding.

© iStock/Thinkstock.

Ancient systems for acquiring and processing materials, such as the stone used to build the Temple of Athena Nike, represent important examples of technological development in their own right.

- The earliest of these stone structures used polygonal stone construction—probably copied from the ruins of those Bronze Age Mycenaean citadels. But over time, Greek masonry construction gradually evolved toward a more refined configuration of rectangular stones set in horizontal **courses** (or layers). This configuration, called **ashlar**, was firmly established by the 5th century B.C.

- Although Greek techniques for quarrying and building in stone were probably borrowed from the Egyptians, the Greeks made substantial improvements to these methods over time. The philosopher Plato acknowledged this point when, in the 4th century B.C., he wrote that the Greeks had invented nothing; rather, they had borrowed all their technologies from other peoples—but then had improved upon everything.

Quarrying Stone

- Greece has always had an abundant supply of **limestone** and **marble** for building. Limestone is a sedimentary rock, usually formed from accumulated skeletal fragments of marine organisms, such as coral. Marble is a metamorphic rock, created when limestone is subjected to intense heat and pressure deep below the earth's surface. Marble is strong, hard, and found in an amazing variety of beautiful colors. It was first incorporated into Greek architecture in the early 6th century B.C. and soon became the preferred material for important public structures.

- Stone for building was supplied largely from two quarries close to Athens: Mount Hymettos and Mount Pentelicon. Such ancient quarries were often located on hillsides, so that blocks of stone could be extracted in a stair-step pattern and then lowered down the hillside for subsequent loading and transport. A hillside location also facilitated the removal of the large quantities of debris that resulted from extracting and shaping stone.

- To extract a block from one of these stair-stepped beds of stone, a vertical trench was cut around all sides of the block. Once all four sides were exposed, the block's bottom surface was freed by chiseling a groove around the base of the block, cutting a series of holes along the groove, and then driving wedges into these holes to split the rock along the horizontal plane.

Transporting Stone

- Well-established quarries, such as Mount Pentelicon, would typically have a track or paved roadway cut into the hillside and used as a ramp to lower stone blocks down the slope. Blocks would be placed on a wooden sled and then guided down this track, controlled by ropes attached to heavy posts called **bollards** mounted at intervals along the descent.
    - At the bottom of the mountain, the block would be loaded onto a barge or oxcart. For overland transport, the largest stone elements required immense wagons, with 12-foot-diameter wheels and pulled by as many as 37 teams of oxen.

- It's worth noting that the ancient road from Mount Pentelicon to Athens followed the contours of the land, allowing these huge vehicles to move along a continual gradual downhill slope.

- The columns of Greek temples were usually built up of cylindrical segments called **drums**—and for the Temple of Artemis, hauling these huge column drums over the 9-mile road from quarry to building site posed a particularly daunting challenge.
  - Chersiphron, the Greek architect who built the monumental Temple of Artemis at Ephesus—one of the Seven Wonders of the Ancient World—responded with a solution that was as simple as it was ingenious: He turned each column drum into a gigantic wheel.

  - The drum was turned on its side, and a simple wooden frame was fitted around it, attached to the drum with a pair of short axles. It then was pulled along by several yoke of oxen.

- With the columns in place, the next major elements required for the temple would have been the **architraves**—rectangular stone beams that spanned across the tops of the columns.
  - The Temple of Artemis architraves were 25 feet long and weighed more than 20 tons each—an even greater transportation challenge than the column drums.

  - Chersiphron's son Metagenes built two 12-foot-diameter wooden wheels and fixed them to the ends of the architrave—which functioned as an enormous axle. He then attached the same sort of wooden frame that his father had used for column drums, hitched up his oxen, and moved the architraves quite successfully.

Positioning the Stone

- On early Greek construction sites, stone blocks were lifted into position by building a temporary earthen ramp and dragging the blocks up the ramp on wooden rollers—just as the Egyptians had done for centuries.

- According to an account by Pliny the Elder, the immense architrave beams of the Temple at Ephesus were raised to their final 60-foot height by this technique—but with a characteristically Greek improvement.

- Chersiphron piled a layer of sandbags on the top of his earth ramp. Then, after each architrave was dragged up onto the sandbags, sand was slowly released from the bags to ease the beam gently down into its final position on top of the columns.

- Building these ramps would have required a huge workforce that few Greek construction contractors could mobilize, however. Eventually, Greek builders developed simple construction cranes that made the task of lifting and positioning stone blocks significantly more efficient. Ingenious Greek engineers solved the problem of lifting blocks without passing a rope underneath.

  - They left stone lugs projecting from sides; a sling was attached to these lugs to lift the block. The lugs were chiseled off later.

  - In some cases, the Greeks carved U-shaped grooves into the ends of the block; a sling was then attached to these grooves to lift the block, and the grooves were covered by adjacent blocks.

  - The most elegant solution was the **Lewis bolt**, a wedge-shaped apparatus that worked so well it is still in use today.

- Mortar was never used in Greek stone masonry construction to position blocks. Instead, each stone was cut to the exact required size, and iron clamps were used to connect adjacent blocks together. Blocks were also sometimes held in place with iron **dowels**. This well-conceived system of iron connectors resulted in excellent structural integrity—particularly appropriate for the many earthquake-prone regions of the Greek world.

Creating the Column

- Although we have used a simple rectangular block to illustrate stone masonry construction, the fabrication process was essentially the same for the many other more elaborate types of building elements used in Greek temples. The column is a particularly interesting example.
    - Greek columns were almost always built up from cylindrical drums stacked on top of each other. These drums were carved roughly to shape in the quarry and then brought progressively closer to their final form in a series of well-planned stages.
    - We actually have documentation of this process in surviving expense accounts from the Erechtheion—another of the fine temples on the Athenian Acropolis.

- Fluted columns were shaped by stripping away layers of stone in four distinct stages. First, the roughly carved outer surface was smoothed to create a perfect cylindrical shape. Next, a thinner layer was stripped off to leave polygonal facets where the fluting would eventually be. The curved fluting itself was hollowed out in two stages, the last after the column was fully assembled—to ensure that each flute formed a smooth, continuous vertical line.

- Like rectangular blocks, column drums also used **anathyrosis**. But for columns, this deliberate hollowing of the mating surfaces was applied to the top and bottom faces rather than the sides. A square socket was carved into the top and bottom of the drum; a hardwood plug was inserted into this socket; and an iron dowel was inserted into a hole drilled into the block. This dowel was used to align adjacent column drums to ensure accurate centering.

An Integrated Technological System

- A masonry wall is the product of a beautifully integrated technological system—one that we cannot begin to appreciate from the completed structure alone. This system required the development of the architect's specifications for each individual stone; synchronization of quarrying, transportation, and construction operations; and careful management of both skilled and unskilled workers: quarrymen, cart drivers, masons, general laborers, blacksmiths (fabricating the iron connectors), and carpenters (building the scaffolding).

- Furthermore, this system had to work efficiently and effectively, because the economic viability of a building project often depended on it. The Parthenon required 50,000 cubic feet of quarried marble—and transporting that stone was the project's single largest expense.
- In the next lecture, we'll see how this technological system produced the extraordinary buildings that have come to symbolize the Hellenic world.

Important Terms

anathyrosis: Hollowing of the end faces of a stone block to facilitate a precise fit.

architrave: A rectangular beam spanning across the tops of two adjacent columns. The architrave is the lowest element of the entablature.

ashlar: A type of masonry construction consisting of rectangular stone blocks set in horizontal rows (or courses).

bollard: A vertical post attached firmly to the ground.

course: A horizontal row of cut stones or bricks.

cyclopean: A type of masonry construction consisting of very large polygonal stones fitted closely together without mortar.

dowel: A cylindrical or rectangular iron peg inserted into the top and bottom surfaces of stone blocks to align and connect them.

drum: A cylindrical segment of a stone column or a cylindrical wall that supports a dome.

Lewis bolt: A wedge-shaped apparatus used to lift stone blocks.

limestone: A sedimentary rock, usually formed from accumulated skeletal fragments of marine organisms, such as coral.

marble: A metamorphic rock, created when limestone is subjected to intense heat and pressure deep below the earth's surface.

Suggested Reading

Malacrino, *Constructing the Ancient World*, chapters 2 and 5.

Oleson, *The Oxford Handbook of Engineering and Technology*, chapter 5.

White, *Greek and Roman Technology*, chapter 7.

Questions to Consider

1. Why did stone masonry construction become so prevalent in the Greek world?

2. How was iron used in Greek ashlar masonry?

3. Explain how the process of quarrying and fabricating stone building components facilitated technological development.

4. Explain how the process of quarrying and fabricating stone building components constituted an important technological system in its own right.

Stone Masonry Perfected—The Greek Temple

Lecture 4

The Classical-era Greek temple was one of the crowning achievements of Hellenic civilization. No other structure of its day could compare in majesty, beauty, or permanence. Functionally, it was simply an enclosure for the cult statue of a deity, but from a spiritual, political, and cultural perspective, it was much more. It was a sacred place where humans might encounter the divine. It was a symbol of the power of the polis—the Greek city-state. As architecture, the temple expressed Greek aesthetic ideals of proportion and symmetry. And to communities throughout the Greek world, it was an icon of Greek cultural identity. As such, the Greek temple was an important milestone in Greek technological development.

Construction of the Greek Temple

- An important example of the Greek temple is the Temple of Concordia, built in the mid-5th century B.C., in Akragas. In a democratic **polis**, such as Akragas, construction of a new temple would normally have been proposed in the popular assembly, perhaps to commemorate some auspicious event, and then voted upon. At this time, temple construction was generally funded from the public treasury.

- Once a temple-building project was approved, a citizen committee was charged with hiring the man who would design the building and supervise its construction. His title in Greek was ***architekton***, a position that also incorporated the responsibilities of the modern structural engineer, construction manager, and master builder.

- Greek temples were typically sited on the acropolis—the highest point of land within the polis—and oriented with the main entrance facing eastward, toward the rising sun.

- Construction of the temple began with excavation for the foundation, which was set on solid bedrock. The foundation was often built of polygonal stone masonry, whose interlocking joints provided better strength and less settling than squared ashlar masonry. The foundation was then backfilled with compacted soil, and above it, a platform called the ***crepidoma*** was constructed of squared ashlar masonry. The highest level, or **stylobate**, formed the temple floor.

© iStock/Thinkstock.

The defining feature of a Greek temple is the colonnade—the row of columns supporting an entablature, running around its perimeter.

The Doric Order

- The columns in the Temple of Concordia are stout and have simple, unadorned **capitals**. These are distinguishing features of the **Doric** architectural style—commonly called the Doric order—which originated in mainland Greece and became popular in Italy and Sicily. The **Ionic** order, which developed a few decades later in Ionia—the Aegean coast of modern Turkey—featured more slender columns and capitals with a distinctive double-scroll motif. The third major order—**Corinthian**—came much later, in the 3rd century B.C., and eventually became the favorite of the Romans.

- In the Doric order, columns were surmounted by a three-level construction called the **entablature**. The lowest level was the architrave—a series of plain stone beams spanning between columns. The second level was the **frieze**. On the exterior of a Doric temple, the frieze consisted of a series of alternating blocks called **triglyphs** and **metopes**. The uppermost layer was the **cornice**, which projected outward to protect the architrave and frieze from the elements.

- On the front of the building, a triangular gable, or **pediment**, was placed above the cornice. The decorative panel within the pediment, called the **tympanon**, was often elaborately sculpted with scenes from mythology or battles.

- The defining feature of a Greek temple was the **colonnade**, a row of columns supporting an entablature. Inside the colonnade was a walled enclosure called the ***cella***, the temple's sacred sanctuary where the cult statue of the god was kept. The east wall contained the main entrance to the sanctuary. The west end had the same configuration but no entrance; this was a purely aesthetic feature that reflected the Greeks' appreciation for symmetry.

- Finally, the building was surmounted by a complex timber roof structure called a **prop-and-lintel** system, covered with terra-cotta or stone tiles.

Origins of the Greek Temple

- The iconic Doric architectural form—with its stepped platform, colonnaded perimeter, and gabled roof—remained essentially unchanged for more than eight centuries. Although there were many variations on the theme, all exhibit the characteristic features of the Doric order—stout **fluted** columns, plain capitals, and a frieze of alternating triglyphs and metopes. The Doric order was the product of three main historical influences:
  - The first of these influences was a primitive wood-frame temple configuration, developed in early-Archaic-period Greece, around the 10th century B.C.

  - The second was the development of terra-cotta roof tile, which significantly influenced temple architecture. Because terra-cotta tiles were much heavier than thatch, they required more robust roof and wall structures, and because these tiles were held in place only by gravity, the slope of the roofline had to be reduced significantly.

  - The third and most important influence was the monumental cut-stone temples of Egypt.

- The result was a wonderful fusion of two architectural styles: The Egyptian stone colonnade was incorporated into the traditional Greek timber temple—and, in a remarkably short period of time, the Doric order emerged. The most intriguing aspect of this synthesis is the extent to which the decorative stone features of the new Doric order preserved the appearance of timber structural elements, called **mutules**, from the primitive wooden temples of the previous century.

A Mathematical Model

- Over time, the Doric and Ionic orders became progressively more standardized. This standardization took the form of an intricate system of geometric design rules for proportioning every element of a temple. This system, as described in detail by Vitruvius, constitutes the first formal system of building design in recorded history.

- In *De Architectura*, Vitruvius tells us that the design of a Doric temple should be based on a single dimensional module equal to the radius of the base of a column in the main colonnade. For example, the height of a column is specified as 14 times the module; a triglyph should be 1 module wide and 1.5 modules high. The height of the cornice is 1/2 module and so on. Another Vitruvius rule governs the number of columns in the colonnade. If there are n columns on the front, then there should be $2n + 1$ columns on each side.

- Using this system, the *architekton* could theoretically design a temple in just three steps: (1) Pick a configuration (e.g., **hexastyle** or **octastyle**), (2) define the module size, and (3) apply the rules to determine all the remaining dimensions of the building.

- This system of empirical design rules was an exceptionally important milestone in the history of technology. In a sense, the system served as a mathematical model of a building. It also fostered the standardization of the Doric and Ionic orders throughout the Greek world, and it provided the equivalent of a modern building code—an enduring record of proven construction standards that could be used by builders.

Entasis: The Optical Refinement

- Although Greek temple architecture was formulaic, in practice, the design rules were not applied rigidly. After all, of the hundreds of Greek temples built, no two were exactly alike. For example, two Doric temples—the Temple of Hera in Paestum and the Parthenon—have very different column proportions. The Parthenon columns are taller and slimmer; the Temple of Hera columns are more sharply tapered, with a pronounced bulge at mid-height.

- That bulge in the Doric column is called **entasis**—one of many optical refinements that characterize Greek architecture. Optical refinements are subtle adjustments designed to create pleasing visual effects.
    - In the Parthenon, for example, none of the building's apparently straight edges is actually straight—creating a "reverse optical illusion." Because the human eye actually perceives a perfectly horizontal line as sagging slightly in the center, the Parthenon's floor (or stylobate) is built with a slight upward arc, as is the entablature.

    - The columns have very subtle entasis, and they all lean slightly inward. The corner columns have a slightly enlarged diameter, because they tend to appear smaller when they are viewed with the sky as a backdrop.

- These optical refinements are purely aesthetic; they have no structural purpose. Nonetheless, from a construction engineering perspective, they reflect extraordinary precision in surveying and stone masonry construction. In the Parthenon, because of all those subtle curves, not a single stone block in the entire structure could be perfectly rectangular. And because such tiny variations in elevation and angle would be lost to even a small amount of foundation settlement, they are also an amazing testament to the Greeks' expertise in foundation design.

Trabeated Construction

- The basic structural system in the Greek temple is **trabeated** construction—a regular arrangement of beams supported on columns. The word "trabeated" derives from the Latin word *trabs*, meaning "beam." This focus on the beams is entirely appropriate because, as it turns out, an inherent structural limitation on the span length of stone beams substantially influenced the design of the Greek temple.

- A column is a vertically oriented structural element that carries load primarily in compression. Because stone is strong in compression, it is well suited for columns. It turns out that in a typical Greek temple, the columns are significantly overdesigned: The compressive stress they experience is much less than the compressive strength of the stone used to build them.
  - It's clear, then, that the temple's robust stone columns are proportioned to meet aesthetic rather than structural criteria.

  - This observation was confirmed by Vitruvius, who tells us that the standard proportions of the Doric column were originally based, not on structural concerns, but on the proportions of the human body.

- A beam is an element that carries load in bending. If we increase a beam's length, width, and height—all in proportion to each other—the stress in the beam increases in proportion to its size. Meanwhile, the strength of stone does not change, regardless of the beam's size. Thus, if we make a series of progressively larger stone beams, eventually, we must reach the point where the stress exceeds the strength—and the beam cannot hold itself up. Simply put, there's a nonnegotiable upper limit on the span of a stone architrave.

- For this reason, trabeated architecture could never be particularly daring. Throughout classical antiquity, the colonnaded temple remained a conservative structural form. Even as it came to represent architectural perfection, the temple's reliance on stone beams prevented it from ever stretching the limits of structural possibility.

- In the next lecture, we'll see this same limitation manifested in a different context—the timber roof structures of temples and similar buildings. And we'll learn about the structural marvel that eventually overcame this limitation—the tie-beam truss.

Important Terms

architekton: Greek term for the man who designed buildings and supervised their construction. (The equivalent Latin term is *architectus*.) The *architekton* served the modern functions of architect, structural engineer, and construction manager.

capital: The decorative top of a column.

cella: Enclosed sanctuary within a Greek temple. The *cella* housed the cult statue of the god to whom the temple was dedicated.

colonnade: A row of columns.

Corinthian order: An architectural style characterized by relatively slender columns and ornate capitals decorated with stylized acanthus leaves.

cornice: The uppermost element of the entablature in a Greek temple. The cornice projects outward to protect the structure from the elements.

crepidoma: The three- or four-step stone platform on which a Greek temple was built.

Doric order: An architectural style characterized by relatively stout columns and simple, unadorned capitals.

entablature: The architectural element spanning across the tops of columns in Greek architecture. The entablature consists of three parts—the architrave, frieze, and cornice.

entasis: Slight bulge in the shape of a Greek column, incorporated to enhance the column's appearance.

fluted: Characterized by vertical grooves carved into the outer surface of a stone column.

frieze: A decorative horizontal band forming one element of the entablature in a Greek temple.

hexastyle: A type of Greek temple with six columns across its front colonnade.

Ionic order: An architectural style characterized by relatively slender columns and double-scroll capitals.

metope: A decorative element in the frieze of a Doric temple. Metopes are always alternated with triglyphs in the Doric frieze.

mutule: A decorative element that represents the ends of angled roof rafters in a Greek temple.

octastyle: A type of Greek temple with eight columns across its front colonnade.

pediment: Triangular gable on the front and rear of a Greek temple.

polis: The Greek city-state.

prop-and-lintel: The timber roof system used in most Greek temples (and other contemporary structures).

stylobate: The top level of the *crepidoma*.

trabeated: A type of structural system consisting of beams supported on columns.

triglyph: A decorative element in the frieze of a Doric temple. Triglyphs are always alternated with metopes in the Doric frieze.

tympanon: Decorative triangular panel within the pediment of a Greek temple.

Suggested Reading

American School of Classical Studies at Athens, *Ancient Athenian Building Methods*.

Coulton, *Ancient Greek Architects at Work.*

Lawrence and Tomlinson, *Greek Architecture.*

Spawforth, *The Complete Greek Temples*.

Questions to Consider

1. How did the decorative stone elements of the Doric order evolve from wooden structural components of Archaic-period temples?

2. Which architectural elements of the Doric order are also structural? Which are not?

3. How was Greek temple architecture influenced by the inherent limitations of stone as a structural material?

From Temple to Basilica—Timber Roof Systems
Lecture 5

Why spend an entire lecture on a roof system? First, the inherent limitations of the prop-and-lintel roof system decisively influenced the form and function of the colonnaded temple and several other principal types of Greek buildings. Second, the subsequent replacement of the prop-and-lintel roof with a fundamentally new structural system—the tie-beam truss—opened new realms of structural possibility and eventually stimulated the development of a significant new architectural form called the basilica. Third, the lack of clear-cut evidence about ancient timber roofs creates a bit of a mystery—and this archeological detective story is really quite fascinating in its own right.

The Prop-and-Lintel Roof

- Not a single wooden component of a Greek temple roof has survived from antiquity. Fortunately, archeological remains of Greek temples have provided us with just enough clues to reconstruct the configuration of timber roof structures with reasonable confidence.

- Using the Temple of Concordia as our example, our primary source of information for understanding the roof system is a series of rectangular sockets carved into the edges of the cornices, pediments, and cross-walls. These sockets provide invaluable clues about the size, position, and orientation of the various roof timbers.

- The ceiling supports would have been substantial timbers because, in addition to their role as load-carrying elements in the roof system, they also supported an attic floor above. The attic of this temple was a functional area—probably a ceremonial space.

- Based on the large sockets in the pediments and cross-walls, we speculate that there were three heavy longitudinal beams. The outer two are called **purlins**, and the one in the center is the **ridge beam**. At the ends of the building, these beams spanned only about 15 feet between the pediments and cross-walls; thus, they were more than adequately supported. Over the *cella*, however, a 60-foot span was much too long; for this reason, there was most likely a series of vertical **props** to provide intermediate support.

- At the ends of the building, the sockets tell us that there should be eight additional purlins supporting the roof over the east and west porches. Finally, closely spaced sockets in the cornices tell us that a series of angled beams, called **rafters**, extended from each cornice across the purlin to the ridge beam. This integrated system of wooden structural elements constituted the prop-and-lintel system. The **lintels** were the ceiling beams, purlins, ridge beam, and rafters.

Roof Tiles

- The prop-and-lintel roof structure supported an ingenious system of interlocking roof tiles, made of either stone or terra-cotta. Lightweight wooden beams called **battens** were on top of the rafters, and large pan tiles were placed across the battens. The bottom row of cover tiles incorporated decorative bosses—called **antefixes**—that hid the exposed edges of the tiles and gave the eaves a finished look.

- Two characteristics of this tile system are particularly noteworthy:
  - First, there were no nails, clamps, or adhesive anywhere in the system; every tile was held in place only by gravity, friction, and mechanical interlocking with adjacent tiles. This is why the slope of Greek temple roofs was relatively shallow; if it were steeper, the tiles could easily slide off.

  - Second, the system is quite waterproof, even though it typically did not include any sort of sealant. Because of the ingenious arrangement of overlaps and upturned edges, the only way for water to penetrate the tile layer would be for it to flow uphill.

Constraints of the Prop-and-Lintel System

- The prop-and-lintel system is another example of trabeated construction. Just as the trabeated stone construction of the temple colonnade was subject to some significant structural limitations, so the trabeated timber construction of the temple roof was similarly constrained.

- The limiting factor in trabeated stone construction is the span length of the architrave beams. If the columns in a colonnade are placed too far apart, the tensile stress in the bottom of the architrave exceeds the relatively low tensile strength of stone, and the beam fails. Of course, this constraint would not necessarily limit the overall size of a Greek temple because the architect could always increase the width of a colonnade by simply adding more columns, rather than by increasing their spacing.

- The difficulty with this approach is that as the front colonnade of the temple gets wider, the *cella* must get wider in the same proportion. And because the *cella* is spanned by the wooden ceiling beams of the prop-and-lintel roof system, increasing the *cella* width soon becomes problematic. As a general rule, Greek builders avoided spanning distances greater than about 30 feet with wooden beams, and the rare exceptions to this rule used truly massive beams.

The designer of the Temple of Hera at Paestum overcame the size limitations imposed by trabeated stone construction by adding more columns to increase the width of the colonnade.

- What are the implications of this 30-foot limitation on wooden beam spans? In the Temple of Hera, the architect provided additional support for the ceiling beams by adding an interior colonnade running longitudinally down the center of the *cella*. This solved the structural problem by cutting the span of the roof beams in half—but it must have also seriously compromised the temple's function by forcing the statue of the goddess to be placed off center.

- Nearly all the later large temples—including the Parthenon—had not one but two interior colonnades. The *cella* walls of the Parthenon were just over 60 feet apart, which it means that it would have been structurally feasible to use a single interior colonnade, as in the Temple of Hera. But instead, the architect chose to use two, allowing the statue of Athena to be appropriately centered in the *cella*.

Invention of the Truss

- We can only marvel at the extent to which architectural function in Greek buildings continued to be compromised by the inherent limitations of wooden beam spans. But everything changed with the revolutionary invention of one of the ancient world's most important structural technologies: the truss.

- The **tie-beam truss** works this way: Start with wooden beam, spanning two colonnades. Add a vertical strut and diagonals, forming the truss. The key improvement over a prop-and-lintel system is that the truss forms a triangle with all corners structurally connected. The structure now carries load—not by bending but in tension and compression. The tie-beam truss is far more efficient than a beam; it can span much longer distances and carry heavier loads.

- There is a great deal of scholarly debate over the origin of the truss but no definitive answer.
  - Several Greek temples on Sicily have *cella* walls that seem too far apart to have been spanned by beams; there's some speculation that Sicilian Greek builders learned about the truss through their frequent contacts with the Carthaginians—a

seafaring people who would have been adept at timber construction. There is even stronger evidence that Hellenistic builders occasionally used trusses.

- But as with so many other ancient technologies, it was the Romans who fully exploited the architectural possibilities of this new structural configuration.

- Thus, as the Greek world gradually succumbed to Roman dominion: The long and narrow **stoa** (located in the **agora** of a Greek city) and the Thersilion-style assembly hall (with its forest of interior columns) were superseded by an architectural form that became possible only with the advent of the truss: the **basilica**.

The Basilica: A New Architectural Form

- The basilica—a new, functionally superior type of public meeting hall—was quintessentially Roman. Located in the forum (or central business district) of a Roman city, it provided a covered space for conducting business and legal transactions. The earliest basilicas were constructed in the late 2nd century B.C.; soon afterward, the form spread widely across the Mediterranean world.

- A variety of distinctly different basilica forms were developed throughout the Roman era, but all had one common distinguishing characteristic—a vast covered central hall, unimpeded by interior columns. Eventually, this same feature led to the basilica's adoption as the architectural model for early Christian churches, after the emperor Constantine converted to Christianity in the 4th century A.D.

- One of the most famous examples of the basilica—and the role of the tie-beam truss in making it possible—is the Roman basilica at Fano, built around 39 B.C. It is the only structure known to have been designed and built by the ancient world's most famous engineer: Marcus Vitruvius Pollio.

- Vitruvius describes the Fano basilica in Book V of *De Architectura*. Unfortunately, his description is maddeningly incomplete, probably because the original treatise included drawings that have not survived. Still, he provides enough detail for us to reconstruct the building's overall dimensions and configuration—including the use of trusses in the roof structure.
  - We know with certainty that the overall plan of the building was T-shaped, with the main entrance centered on the long side of one wall and with the stem of the T serving as the tribunal—an area with a raised floor, where the local magistrates heard cases and dispensed justice.

  - We know, too, that the heart of the building was a vast central hall—120 feet by 60 feet—outlined by 50-foot monolithic stone columns, surrounded by a two-story aisle.

  - The aisle had a lower roofline than the hall to allow for illumination by a row of windows on the upper level. And the roof of that monumental central hall was supported by a series of 60-foot wooden tie-beam trusses.

  - These trusses made the basilica possible. Without them, the only way to support the roof would have been with a forest of columns filling the central hall. But this would have fundamentally altered the architectural design, robbing the basilica of its most characteristic feature: unimpeded open space.

- Vitruvius's basilica represented a monumental leap forward in both structural engineering and construction technology.
  - The fact that the Romans could erect 50-foot monolithic columns and 60-foot timber trusses in a routine, small-town construction project suggests that they brought a new spirit to the engineering enterprise—a spirit of innovation, informed by a strong appreciation for structural principles and greatly enhanced by a cultural predisposition toward pragmatism and organization. Here, we see the roots of a construction

revolution—one that will reach its climax in the grand public works of imperial Rome.

- In the next lecture, we will explore the building blocks of that revolution—the two most important construction technologies of the ancient world: concrete and the arch.

Important Terms

agora: The central marketplace of a Greek city.

antefix: A decorative boss used to hide the exposed ends of roof tiles along the lower edge of a roof.

basilica: A Roman public building characterized by a large covered central hall.

batten: A lightweight beam that directly supports a row of roof tiles.

lintel: A horizontal structural element that spans an opening, such as a door or window.

prop: A vertical strut forming one element of the prop-and-lintel roof system.

purlin: A longitudinal beam that supports the rafters in a prop-and-lintel roof system.

rafter: An angled beam that supports a roof.

ridge beam: A longitudinal beam that supports the peak of a roof.

stoa: A long, narrow Greek building used to house shops and offices, usually located in the agora.

tie-beam truss: A structural system composed of members configured in interconnected triangles.

Suggested Reading

Hodge, *The Woodwork of Greek Roofs*.

Ulrich, *Roman Woodworking*.

Vitruvius, *The Ten Books on Architecture*, Book V.

White, *Greek and Roman Technology*, chapter 7.

Questions to Consider

1. How did the inherent limitations of the timber prop-and-lintel roof system influence the architectural design of Greek buildings?

2. How does a tie-beam truss differ structurally from a prop-and-lintel roof?

3. How did the tie-beam truss facilitate the development of the Roman basilica as an architectural form?

Construction Revolution—Arches and Concrete

Lecture 6

Roman engineers sought to overcome the limitations of both trabeated stone construction and the timber truss with a fundamentally new structural form—one that allowed for both long spans and permanence. By doing so, they created a form that would become intimately associated with Roman civilization itself: the arch. Roman engineers then began experimenting with the arch as a means of enclosing space—an architectural purpose that required a new structural configuration: the vault. However, the full potential of the vault could not be realized as long it had to be meticulously assembled from geometrically complex wedges of stone. The Roman construction revolution needed an innovative, more versatile material that could substitute for stone: concrete.

The Arch

- The **arch** was not a Roman invention; archeologists have found many examples that predate the Romans—in Israel, Egypt, and Greece. But the arch was not fully realized as a structural technology until it was embraced by the Romans. As with the truss, Roman engineers saw immense potential in the arch where other civilizations had not, and they exploited that potential in varied and innovative ways.

- By the Roman imperial period, the arch had become ubiquitous. In the hands of the pragmatic, technologically adept Romans, the arch led to unprecedented structural and architectural innovation. We see it in bridges, city gates, aqueducts, triumphal arches, sewer systems, and great public buildings, such as baths, basilicas, and arenas.

- A key feature of the arch is its ability to carry load entirely in compression. The arch is capable of supporting itself only after complete; thus, the Romans used centering, added wedge-shaped **voussoirs**, and put the keystone in last. Because it carries load in compression, the arch effectively overcomes the principal limitation of trabeated stone construction—short beam spans, resulting from the low tensile strength of stone. The arch can be made of many small stones rather than a single large one, and dramatically longer spans are possible.

- An arch, however, can only carry load successfully if it is supported laterally. The weight of the arch and any additional superimposed loading causes it to flatten; its ends move outward. This tendency is called **thrust**. To carry load, this outward thrust must be resisted by inward lateral support.

- Thus, the arch's unique advantage (ability to carry load entirely in compression) facilitated its use in Rome's greatest and most innovative structures, yet its unique disadvantage (need for lateral support) profoundly influenced both the form and function of these structures.

The Vault

- The earliest Roman arches were typically used in settings where lateral thrust was easily restrained, whether by a surrounding wall, compacted soil, or foundations. An arrangement of multiple adjacent arches is called an **arcade**. Early applications of the arch, such as in gateways and bridges, were quite effective in spanning horizontal distances. But the most revolutionary applications were yet to come.

- In the 2nd century B.C., Roman engineers began experimenting with the arch as a means of enclosing space, which required a new structural configuration: the **vault**. The simplest type of vault, called the **barrel vault**, was created by simply extruding an arch into the third dimension to create a curved roof or ceiling.

- But while the vault provided a versatile new tool for enclosing space, its potential could not be fully realized as long it had to be meticulously assembled from geometrically complex wedges of stone. The Roman construction revolution needed a new, more versatile material that could substitute for stone: **concrete**.

Early Roman Building Methods

- Early Romans most likely learned stone masonry construction from their Etruscan neighbors. Because Etruscan architecture had been strongly influenced by the Greeks, Rome's early monumental structures used essentially the same Greek-style ashlar masonry construction: precisely cut stone blocks, assembled in horizontal courses without **mortar**, interconnected with iron clamps. The Romans called this configuration ***opus quadratum***, Latin for "squared work."

- Initially, the only significant difference between Roman *opus quadratum* and Greek ashlar masonry was the stone itself. During the Roman Republic, Roman builders had little access to marble, so they built primarily with **tufa**, a locally available limestone. Tufa was often faced with a layer of stucco and decorated to resemble marble—in imitation of Greek masonry.

- Over time, however, the Romans developed some significant improvements on Greek-style masonry. Most important, they began using a thin layer of **lime mortar** in the joints between stones.

Concrete

- Although lime-based mortar works quite effectively when used in thin layers to create joints between stones, it could not be used effectively as a substitute for stone—as concrete eventually was. If lime-based mortar is cast in a thick mass, like concrete, its outside surface hardens through direct exposure to the air, but because this crust then prevents air from penetrating inside the mass, the inner core never fully hardens.

- This significant limitation was overcome when the Romans discovered a naturally occurring volcanic ash called **pozzolana**. The unique chemical properties of pozzolana made Roman concrete possible. When combined with lime and water, pozzolana undergoes a chemical reaction to produce a substance five times stronger than lime-based mortar. More important, this chemical reaction does not require air; thus, pozzolana-based mortar can be cast in large masses without hindering the curing process. It will even harden underwater.

- Eventually, builders began experimenting with true concrete, created by mixing liquid pozzolana mortar with solid material, or **aggregate**, for added strength. In modern concrete, the aggregate is a carefully proportioned mix of sand and gravel; in Roman concrete, it was usually random rubble: fist-sized stones, broken bricks, pieces of terra-cotta tiles. When the liquid mortar and aggregate were mixed together, the aggregate formed an interlocking matrix of solid material, with smaller pieces filling the spaces between the larger ones. The mortar paste then filled the smallest voids and hardened, creating a dense, rocklike mass with compressive strength comparable to stone.

Mass Production and Standardization

- Roman concrete first came into widespread use in a series of progressively more sophisticated systems for wall construction. The first was called ***opus incertum***. In this system, two outer wall facings were built of random, fitted stones, and the space between the facings was filled with concrete. A Roman temple at Terracina has the distinctive appearance of *opus incertum* in the wall. There are rectangular-cut stones for reinforcement at the corners. (The architectural term for this detail is a **quoin**.)

- By the end of the 2nd century B.C., builders had refined this system to create ***opus reticulatum***, which retained the solid concrete core of *opus incertum* but replaced the random stone facings with pyramid-shaped cut-stone blocks arranged in a diamond pattern. *Opus reticulatum* represented a significant improvement over *opus incertum* because the pyramid-shaped blocks could be mass produced in standard sizes by unskilled workers.

- This trend toward mass production and standardization accelerated further with the development during the imperial era of an even more effective system called ***opus testaceum***. Because of its many advantages, *opus testaceum* quickly became the dominant construction method throughout the Roman Empire—except in the Hellenized east, where traditional stone masonry retained its popularity.

Rome as an Imperial Power

- The widespread use of *opus testaceum* was directly tied to Rome's growing role as an imperial superpower. Rome's imperial ambitions demanded a vast increase in all forms of construction: city walls, bridges, aqueducts, basilicas, baths, palaces, apartment complexes. At the same time, wars of conquest produced large numbers of low-skilled slave laborers.

- Traditional masonry construction relied heavily on highly skilled masons for cutting and fitting stone. But because *opus testaceum* used only standardized mass-produced materials, assembled in an unvarying pattern, the system could be implemented with gangs of unskilled workers, supervised by just a few skilled masons. Thanks to *opus testaceum*, construction became more a matter of assembly than of craft.

- Yet this humble construction method lies at the heart of innumerable grand imperial projects of mind-boggling scale and complexity. We'll explore many of these—the Colosseum, Nero's Domus Aurea, Trajan's Market, the Pantheon, the Baths of Caracalla—in future lectures.

- The widespread adoption of *opus testaceum* stimulated a tremendous expansion of the Roman brick-making industry around the 1st century A.D. Brick factories around the empire mass-produced millions of terra-cotta bricks in standard sizes—ranging from the 8-inch-square **bessalis** to the 2-foot-square **bipedalis**. The triangular bricks used in *opus testaceum* were then cut from these square bricks on the job site.

- Roman brick makers soon expanded their product lines to include a variety of other standardized terra-cotta elements: hexagonal floor tiles, circular and wedge-shaped bricks for building columns, water pipes, and rectangular ducts called tubuli, which were used in heating systems.

Two Revolutionary Technologies

- Two revolutionary technologies—the arch and concrete—converged most powerfully in the concrete vault, which was perhaps the most distinguishing feature of imperial Roman architecture. The simple barrel vault could be constructed of cut stone, albeit with considerable difficulty. But with concrete, vaults of much greater complexity could be built quite easily. A **groin vault** was formed by two barrel vaults intersecting each other perpendicularly.

- The most important characteristic of concrete is its versatility of form. Any shape that can be modeled as wooden formwork can be transformed into a strong, waterproof, monolithic architectural element by simply filling that form with concrete. Using this

One of the earliest known uses of the *opus incertum* technique is seen in the Porticus Aemilia, ancient Rome's largest commercial building.

technique, complex vaulted forms—and even domes—could be created with relative ease. When the Romans figured out how to combine concrete, arches, and vaults in integrated structural systems, the door was opened to the grand public works of imperial Rome.

- In those beautifully integrated concrete arches and vaults, we can see a potential that was fully realized two centuries later in Rome's grandest building: the Colosseum. We will explore that magnificent structure in the next lecture.

Important Terms

aggregate: Sand, stone, or rubble used as a component of concrete.

arcade: An arrangement of multiple adjacent arches.

arch: A structural element that can span a horizontal distance while carrying load primarily in compression.

barrel vault: A vault with the shape of a half-cylinder.

bessalis: An 8-inch-square Roman brick.

bipedalis: A 2-foot-square Roman brick.

concrete: A manufactured structural material created by combining cement, water, and aggregate.

groin vault: A vault formed by the intersection of two perpendicular barrel vaults.

lime mortar: A type of mortar manufactured by baking limestone in a kiln.

mortar: A substance used to fill the gaps between stones or bricks in masonry construction.

opus incertum: A Roman wall construction system, consisting of outer facings of random fitted stones surrounding a concrete core.

opus quadratum: Roman term for ashlar stone construction.

opus recticulatum: A Roman wall construction system, consisting of outer facings of pyramid-shaped stones surrounding a concrete core.

opus testaceum: A Roman wall construction system, consisting of outer facings of overlapping triangular bricks surrounding a concrete core.

pozzolana: A naturally occurring volcanic ash used as the cement in Roman concrete.

quoin: A rectangular stone used to reinforce the corner of a building.

thrust: The outward force generated by an arch under load; also, the tendency of an arch to spread out laterally under load.

tufa: A soft, porous limestone commonly used in Roman construction.

vault: An arched roof or ceiling.

voussoir: A wedge-shaped component of an arch.

Suggested Reading

Malacrino, *Constructing the Ancient World*, chapter 6.

Oleson, *The Oxford Handbook of Engineering and Technology*, chapter 10.

Taylor, *Roman Builders*, chapter 3.

White, *Greek and Roman Technology*, chapter 7.

Questions to Consider

1. What are the principal advantages and disadvantages of the arch as a structural element?

2. How did the need to restrain the lateral thrust of arches and vaults influence the architectural design of Roman buildings?

3. Why did concrete largely supersede stone as a structural material in the Roman imperial period?

Construction in Transition—The Colosseum

Lecture 7

Just as the colonnaded temple is the iconic symbol of Classical Greece, so the Colosseum has become emblematic of imperial Rome. Its immense size reflects the vast, encircling reach of the empire. Its purpose reminds us of the love of spectacle that was so much a part of Roman life. The Colosseum's architectural layout is a monument to Roman organizational genius. And Roman philhellenism—love of all things Greek—is evident in the incorporation of all three Greek architectural orders in the building's facade. The Colosseum is also singularly important as a snapshot of Roman engineering at a pivotal time of transition—a time when the Roman construction revolution was manifesting itself in radically new approaches to building design.

Unprecedented in Scale

- From the very beginning of the Roman Empire, in 27 B.C., Roman emperors saw the need to sponsor grand public works—to win the favor of the masses, to demonstrate the empire's power, and to glorify themselves. Without question, the grandest manifestation of this epic building boom was the Colosseum.

- The Colosseum is more properly called the Flavian Amphitheater, because it was built under the patronage of all three emperors of the Flavian dynasty. The project was initiated by Vespasian, who placed his great arena in the very heart of Rome—on a piece of land where his much-despised predecessor Nero had built an artificial lake to adorn the grounds of his opulent private residence: the Domus Aurea, or Golden House.

- The uppermost level was completed by Vespasian's son and successor, Titus, in A.D. 80. After Titus died the following year, Vespasian's other son Domitian added the ***hypogeum***—a two-story network of corridors, rooms, ramps, and shafts located underneath the wooden arena floor.

- The Colosseum was unprecedented in scale. The oval-shaped structure was 620 feet long by 510 feet wide and as tall as a modern 16-story building. Its seating area—called the ***cavea***—could accommodate 55,000 people. The arena floor was only a bit smaller than a modern American football field. Its outer wall alone required 130,000 cubic yards of stone and was held together with 300 tons of iron clamps. That the bulk of this extraordinary structure could have been built in just five years is nothing short of astonishing.

A Blend of Tradition and Innovation

- The Colosseum reflects a transitional period in Roman architecture history and is an embodiment of the Roman construction revolution.
    - Its monumental outer wall and principal interior load-carrying columns were built entirely of traditional *opus quadratum*; yet many of its interior walls were of *opus testaceum*, an innovative composite wall system consisting of brick facings enclosing a solid concrete core. All its vaults are state-of-the-art cast concrete—formed into an amazing variety of complex configurations, dictated by the geometric demands of the building's oval shape.

    - As such, the Colosseum represents a grand compromise—a characteristically Roman blend of tradition and innovation.

- The plan of the Colosseum is often described as an **ellipse**. Although there is still some scholarly debate on this subject, the shape is almost certainly not a true ellipse in the strict mathematical sense. Rather than using a mathematically correct ellipse, Roman engineers laid out the plan of the Colosseum as a series of interconnected circular arcs, each with a different radius and center point.

- Because of its multiple centers, this shape is called a **polycentric oval**. Because each segment of the oval is actually a circular arc, the engineer could lay out perfectly parallel corridors quite easily by just drawing concentric arcs. Radial corridors were equally easy to lay out by drawing straight lines outward from the center of the associated arc.

- We can see the clear advantage of the polycentric oval. Unlike the elliptical layout, a grid of circular arcs and radial lines produces a regular array of modules, each with four perfectly square corners. This geometric configuration was quite advantageous, not only for simplicity of layout, but also for stonecutting.

Built for Permanence

- The Colosseum has a two-level foundation. The lower level is an oval-shaped ring of ***opus caementicium***—solid concrete, 23 feet thick and just slightly longer and wider than the building's eventual footprint. The second layer, nearly as thick, was created by first enclosing its perimeter with brick walls and then filling the resulting enclosures with concrete. The upper surface of this layer is the building's ground floor.

- On this solid working surface, the builders carefully laid out the first-level floor plan, consisting of 80 radial corridors delineated by seven rings of heavy rectangular columns called piers.
    - Certain piers were made of a strong, hard stone called **travertine**; the interconnecting walls were made of much softer, weaker tufa. Because of the difference in strength, it is clear that the travertine piers are the principal load-carrying elements supporting the structure above, and the tufa walls serve only to enclose the corridors.

    - Some piers incorporated an **engaged column**—a purely decorative element consisting of a half-column projecting from a wall or pier.

- The huge stone blocks used to build the walls could only have been lifted and positioned with construction cranes. The standard Roman crane was an improved version of the Greek crane. Because of its inverted-V configuration, it could pivot forward and backward quite easily but could only be rotated from side to side with great difficulty. Thus, the only practical way to build closely spaced radial walls was to position one crane on the outside and one on the inside, so that a stone block could be placed at any point on the wall simply by pivoting one of the cranes forward or back.

Maximizing the Versatility of Concrete

- With the stone elements of the Colosseum's first level in place, construction of the vaulting began. Every radial corridor, **annular corridor**, and stairway required a concrete barrel vault. In the annular corridors, vaults were constructed on wooden centering supported on cornices above the arcades. This arrangement was particularly efficient, because it required no temporary supports below and, thus, allowed workers to continue using the corridors, even while concrete was being poured. Above the centering, brick ribs were installed at intervals so that sections of vaulting could be constructed in separate concrete pours of manageable size.

- Three of every four radial corridors housed a concrete stairway. To build these, an angled shelf was first cut into the side walls at the proper angle, then centering was constructed as a series of semicircular frames covered with wooden boards. Finally, concrete was poured in successive lifts to create the steps.

- Vaulting inclined on an angle is called a **raking vault**. The geometry of a raking vault in a wedge-shaped corridor is quite complex, yet building it was not appreciably more difficult than building a simple barrel vault because its shape is fully defined by a series of simple, semicircular centering frames. This is a dramatic demonstration of the versatility of concrete.

Use of *Opus Testaceum*

- On the first level of the Colosseum, the walls were constructed entirely of *opus testaceum*. These inner walls supported only the lower tier of the *cavea* on short-span raking vaults; thus, they were relatively lightly loaded. But the main piers supported one or more upper floors in addition to the *cavea*, which meant their loads were much higher.

- The second-level structure was largely the same as the first level, with travertine arcades forming two outer annular corridors, covered with concrete vaulting. The upper surfaces of these *cavea* vaults were flat. The actual rows of seating were carved from marble blocks and installed later. The only major difference between the second-level and first-level structure is that *opus testaceum* was used in place of tufa for the in-fill walls between the piers.

- On the third level, travertine piers extended upward to support the arcades of an outer annular corridor and an *opus testaceum* wall formed a lower inner corridor—covered with an annular vault that also shaped the top tier of the *cavea.*

A Significant Structural Innovation: The Flying Stairway

- Within the inner corridor of the third level was a seemingly minor feature that was actually one of the Colosseum's most significant structural innovations. On the first two levels, all the stairways were oriented radially. But because the third level is composed entirely of two annular corridors, conventional radial stairways would not work here; they would have completely blocked at least one of the corridors, severely limiting circulation.

- The designer overcame this constraint quite ingeniously—by using a pair of narrow circumferential stairways leading to an intermediate landing and then spanning across the corridor overhead to access the *cavea.* This flying stairway was unprecedented in ancient architecture—a small but daring structure that speaks volumes about Roman engineering savvy.

A Series of Engineering Challenges

- The uppermost level of the Colosseum was a flat floor that supported wooden bleachers—the proverbial "cheap seats" reserved for women and slaves. Here, the outer wall was no longer an open arcade but, rather, a solid wall of travertine, thickened and strengthened with *opus testaceum* on the inside. The inner edge of this level was ringed with a Corinthian colonnade supporting a

Rome's hierarchical social structure was literally "set in stone" in the Colosseum's rigidly segregated *cavea*, from the emperor's viewing box at the lowest level to the bleachers for women and slaves at the highest.

wooden roof structure—a throwback to the trabeated construction of Classical Greece.

- Although the structure on this level was relatively simple, erecting it must have been one of the greatest challenges in the entire project. The 20-foot stone columns weighed about 10 tons each. Hoisting them 19 stories upward and then maneuvering them into position would have required skillful use of high-capacity cranes—yet there was precious little room for such devices on the narrow floor. And once the colonnade was in place, the working space for maneuvering travertine blocks for the outer wall would have been even more restricted.

- Also on the fourth and highest level are the mountings for heavy wooden masts that supported the Colosseum's most unusual and innovative feature—the rope-and-canvas awning called the ***velarium*** that shaded the audience members from the sun as they enjoyed the spectacle. Just the rope and canvas necessary to shade the *cavea* would have weighed about 150,000 pounds. Supporting that load across a 600-foot span would have been quite a challenge—particularly because the tension in the supporting ropes would have generated huge inward forces on those wooden masts.

A Surprisingly Conservative Design

- The Colosseum was—and is—a magnificent structure. Yet with the exception of the *velarium* and the unique flying stairways, its structural design is surprisingly conservative. None of its arches exceeded 20 feet in span, even though spans five times longer were common during this era. Its main structural columns were tried-and-true travertine, while more modern *opus testaceum* was used only in lightly loaded walls.

- It is quite likely that the Colosseum's engineers made these conservative decisions precisely because they were designing a building of such unprecedented scale. Thus, even in their conservatism, we can see a sophistication of judgment that modern engineers would do well to emulate.

- The Colosseum represents an important milestone along the road to greater confidence in the use of *opus testaceum* and concrete. Fully developed, that confidence would manifest itself in an extraordinary flowering of structural innovation in the great vaulted buildings of imperial Rome. We'll explore these in the next lecture.

Important Terms

annular corridor: A ring-shaped corridor that runs around the circumference of a circular or elliptical structure.

cavea: The seating area of a Roman amphitheater.

ellipse: The geometric shape formed by a set of points that are the same total distance from two points, called foci.

engaged column: A decorative half-column projecting from a wall or pier.

hypogeum: A two-story network of corridors, rooms, ramps, and shafts located underneath the wooden arena floor of the Colosseum.

opus caementicium: Roman term for solid concrete.

polycentric oval: The geometric shape formed by a series of interconnected circular arcs.

raking vault: A barrel vault inclined at an angle.

travertine: A strong, hard limestone commonly used in Roman construction.

velarium: The rope-and-canvas awning used to shade spectators at Roman amphitheaters.

Suggested Reading

Malacrino, *Constructing the Ancient World*, chapter 6.

Taylor, *Roman Builders*, chapter 4.

Questions to Consider

1. In designing the Colosseum, why did Roman engineers continue to use traditional Greek-style stone masonry even though the superior *opus testaceum* system was already well-established?

2. What is the structural advantage in arranging arches around the circumference of a circular or oval-shaped building?

3. How did the limitations of positioning construction cranes influence the design of the Colosseum?

4. What was the role of Roman emperors in fostering technological development?

The Genesis of a New Imperial Architecture
Lecture 8

Completed in 193 B.C., the Porticus Aemilia, with its sophisticated arrangement of concrete arcades and stepped vaults, was one of the earliest expressions of the Roman construction revolution. Surprisingly, though, its innovative design did not prompt widespread adoption of vaulted concrete construction. That would come in the 1st century A.D., with an explosion of vaulted concrete construction, much of it exhibiting bold approaches to enclosing interior space. This lecture examines the causes and effects of a seismic shift in Roman architecture, focusing on two milestone structures: Nero's Golden House and Trajan's Market. We'll explore how a Roman architectural revolution was initiated by a confluence of new technologies, human genius, and a horrific catastrophe.

An Architectural Perfect Storm

- A coherent, fully developed vaulted architecture did not emerge until A.D. 54, when the emperor Nero came to power. Nero, a nonconformist and aspiring artist, immediately initiated a period of intense architectural experimentation. At this pivotal moment in history, three preconditions for the full flowering of a bold new Roman architecture were finally in place:
  - First, Roman architects and engineers had acquired proficiency and confidence in the use of *opus testaceum* and cast-concrete vaulting. A full 250 years after the Porticus Aemilia, the technology of the Roman construction revolution was finally ripe for exploitation.
  - Second, large-scale civil works projects already sponsored by the emperors had stimulated tremendous growth in construction capacity—brick-making factories, quarries, experienced construction contractors, and craftsmen.

  - Most important, absolute power was now in the hands of an artistically inclined emperor who consciously sought to break with tradition—and was inclined toward grandiose gestures.

- With these conditions in place, a period of architectural innovation was inevitable—though it might have taken time. But, as it turns out, a cataclysmic event would cause the revolution to happen far more dramatically than even Nero might have imagined.

A Cataclysmic Event

- On the night of July 18, A.D. 64, fire broke out in the area between Rome's Palatine and Caelian hills and was driven across the city by strong winds. The fire burned for nine days, leaving most of Rome a smoldering ruin. Nero ordered the city rebuilt—not as it was before but according to a new, rational plan, organized around broad, straight avenues interspersed with open plazas. Nero decreed that all reconstructed buildings would be limited in height and built of fire-resistant materials. At his own expense, he added porticoes to the fronts of the city's apartment buildings to provide elevated platforms that could be used by firefighters in future blazes.

- The great fire, and Nero's response to it, transformed Rome. Almost immediately, the entire city became one vast construction site—a blank canvas for creative architects and engineers looking for opportunities to experiment with new ideas. And to meet the need for architectural flexibility, rapid construction, and fire resistance, there could be no better raw materials than *opus testaceum* and concrete—which could be assembled by gangs of unskilled laborers using standardized, mass-produced components. From A.D. 64 onward, the structural fabric of Rome—and much of the wider empire—would be brick and concrete.

- The fire had another far-reaching consequence: It destroyed Nero's Domus Transitoria and sparked his interest in building an even grander palace. Thus was born the Domus Aurea, or Golden House—one of the most extravagant private residences ever built. The ruins of the surviving wing of the Domus Aurea are sufficiently intact to convince us that its creators, Severus and Celer, succeeded in constructing a sophisticated system of vaulted spaces quite unlike anything that had preceded it.

The Domus Aurea Atrium

- The Domus Aurea atrium was arguably the most innovative structure in ancient Rome. In its central feature, an octagonal room, was something never before seen: flat lintels made entirely of brick. But these lintels are not beams at all; they are arches. Because they are not curved, they are often called **flat arches**.

- In the Domus Aurea, the **impost blocks** are made of travertine—very hard, strong limestone. Their angled ends serve as supports and are critical to the functioning of the arch.

- The flat arch carries load in exactly the same way a conventional curved arch does. Indeed, the path of the internal compression force in the flat arch actually follows a curved path from support to support. The curve always remains perpendicular to the joints between bricks, allowing transmission of the compressive force from brick to brick.

The Domus Aurea Dome

- The octagonal walls of the atrium transitioned seamlessly to a cast-concrete dome, which was pierced by a circular opening, or **oculus**, on top. The oculus was formed by a ring of radially oriented bricks. A concrete dome carries load much like an arch. When the dome is placed on top of the atrium's octagonal walls, the thrust of the dome causes the walls to tip outward, resulting in even greater instability.

- Severus and Celer met this challenge in two ways. First, they added heavy piers, called **buttresses**, at the corners of the octagon, to prevent the walls from tipping outward. Second, the designers further stabilized the dome by providing vaulted chambers extending outward from five sides of the octagon. Because these chambers were oriented perpendicular to the atrium walls, they were effective in preventing the walls from tipping outward. For this reason, they are often called **buttress vaults**.

- Another fascinating aspect of the architectural design is that the octagonal space around the outside of the dome was left open to create a skylight, which opened into the vaulted chambers through short vaults. Because they were arranged around the octagon, these "light wells" provided excellent natural illumination, regardless of the sun's position in the sky.

- The atrium of Nero's Domus Aurea was an extraordinary architectural statement. In describing this milestone structure, J. B. Ward-Perkins, a noted architectural historian, wrote, "It is hardly an exaggeration to say that the whole subsequent history of European architectural thought hangs upon this historic event."

The World's Oldest Shopping Mall

- If Nero's Domus Aurea was the seed of a new architectural paradigm, then its full flowering can be seen in one of Rome's most underappreciated buildings: Trajan's Market. Emperor Trajan, who ruled from A.D. 98 to 117, was a prolific builder. He was aided by a gifted architect-engineer named Apollodorus of Damascus, a Greek from Roman Syria.

- Trajan's Market was built around A.D. 106 to 112, and much of it has survived—probably because it was repurposed as a religious establishment during the Middle Ages. Trajan's Market was a magnificent architectural design, brilliantly adapted to a challenging site. A well-integrated arrangement of shops, multistory halls, corridors, streets, ramps, stairwells, and terraces, the market provided shoppers with easy access, effective illumination, protection from the elements, and splendid views.

- The market's design fully reflected the new imperial architecture, enabled by the stuff of the Roman construction revolution—*opus testaceum* and concrete, with only minimal stone—and not a single structural column.

© iStock/Thinkstock.

Today, the ruins of Trajan's Market include some 170 rooms distributed among six levels, arrayed up the slope of the Quirinal Hill to a height of 115 feet.

Trajan's Market

- Trajan's Market was delineated by a stone **hemicycle** that incorporated 13 wedge-shaped rooms covered with concrete barrel vaults. Eleven of these were shops (called *tabernae*), and the other two were stairwells. The hemicycle was flanked by three large semicircular halls.

- On the second level, these three halls were covered with concrete half-domes, or **semi-domes**, each creating a majestic space that, unlike a full dome, was illuminated by second-story windows piercing the front wall. One level up, the hemicycle was ringed by a promenade overlooking the Forum, and a third level of *tabernae* opening in the opposite direction.

- At the northeast corner of the upper structure was the most unique and innovative component of Trajan's Market: a cavernous two-story hall, called the *aula* (Latin for "courtyard"). Each side of the aula was defined by blocks of *tabernae*, stacked on two levels, with the upper-level entrances set back to create a corridor. Overhead, resting on stout piers, was an exceptionally fine example of early Roman groin vaulting.

Groin Vaulting and the Flying Buttress

- Apollodorus did not invent groin vaults. They had been used—sparingly and conservatively—for at least 200 years before his time. What Apollodorus did was to use them in a new and daring way—supported, not on solid walls, but on piers to create a dramatic covered space, brilliantly illuminated by natural light. In the *aula* of Trajan's Market, we see an architectural concept that, 1,000 years later, would be manifested in the grand cathedrals of the Middle Ages.

- Beyond its striking visual effect, Apollodorus's groin vaulting was also an amazing technological achievement. Like any arched structural element, groin vaults generate significant lateral thrust, and when they are supported on tall piers, this thrust causes a strong tendency to tip the piers outward. Apollodorus responded to this challenge with two innovations that demonstrate his mastery of structural engineering:

    - First, he supported the vaults on stone projections called corbels. Because they projected inward, the corbels significantly reduced the vaults' tendency to tip the piers outward.

    - Second, Apollodorus arranged the *tabernae* along the sides of the main hall, so they would function as buttress vaults to prevent the piers from tipping outward. He added ingenious brick arches, which provide a load path for transmission of lateral thrust outward from the groin vaulting across the corridor and into the upper-story buttress vaults. These arches represent an astonishing structural innovation for which Apollodorus of Damascus is seldom given credit: the **flying buttress**.

A New Paradigm

- In both Nero's Domus Aurea and Trajan's Market, we see the advent of a fundamentally new imperial Roman architecture.
    - Traditional trabeated construction was abandoned for complex vaulted structural systems with few, if any, columns.

    - Symmetrical, rectilinear plans were replaced by circular, polygonal, and irregular geometric forms.

    - Exterior appearance was deemphasized in favor of interior spaces that were visually striking, cleverly illuminated, and highly functional.

    - And while these new architectural forms were enabled by technology—*opus testaceum*, pozzolana-based concrete, and ever-improving construction processes—they were ultimately the product of human aspirations and ingenuity.

- This new Roman architecture is appropriately characterized as imperial, because it was invariably sponsored by the emperors, but we really owe it to the genius of such men as Severus, Celer, and Apollodorus, who had the vision to imagine a fundamentally new archetype and the skill to make this vision a reality.

- In the next lecture, we'll immerse ourselves in ancient Rome's most notable structural engineering achievement: the Pantheon.

Important Terms

buttress: A pier or thickened section of a wall that resists the lateral thrust of a vault or dome.

buttress vault: A barrel vault used to restrain the lateral thrust of a larger vault, arch, or dome.

flat arch: A flat, horizontal structural element that carries load as an arch; also called a lintel arch.

flying buttress: A structural element that resists the lateral thrust of ceiling vaults. Flying buttresses are external to the structure they support.

hemicycle: A semicircular architectural feature.

impost block: A stone block that supports the base of an arch.

oculus: The circular opening at the top of a dome.

semi-dome: A half-dome.

Suggested Reading

MacDonald, *The Architecture of the Roman Empire*, chapters 2 and 4.

Taylor, *Roman Builders*, chapter 5.

Questions to Consider

1. What personal qualities of Emperor Nero helped foster the Roman construction revolution?

2. What characteristics of Nero's Domus Aurea reflect the advent of a new architectural paradigm?

3. What characteristics of Trajan's Market reflect that new architectural paradigm in its fully mature form?

4. How was the development of this new imperial architecture influenced by technology?

The Most Celebrated Edifice—The Pantheon

Lecture 9

As a work of architecture and engineering, the Pantheon has been often imitated but never equaled. It was the ancient world's most ambitious engineering achievement—and the quintessential example of the architectural style of imperial Rome. Completed in A.D. 126, during the reign of Emperor Hadrian, the Pantheon is a temple dedicated to all the gods. It is the best preserved of all ancient Roman buildings; thus, it provides us with a rare opportunity to experience imperial architecture as the Roman citizens of that era did.

Emblematic of Imperial Power

- Everywhere we look, outside and in, the Pantheon proclaims the glory of Rome. Although the front porch, or **portico**, of the Pantheon bore a striking resemblance to a Greek colonnaded temple, several characteristically Roman features prove that this was not just a copy of a traditional Greek building. The typical Greek temple sat on a stepped base, or stereobate, accessible from all four sides. In contrast, the typical Roman temple was built on a taller platform accessible only from the front; the Pantheon followed this Roman convention.

- In addition to the portico's frontal orientation, the Pantheon's columns were also decidedly "un-Greek" in character. The capitals were in the typically Roman Corinthian style, and the column shafts were not fluted, as Greek columns would have been.
  - Most important, rather than being built up from a series of cylindrical drums, each of the Pantheon's column shafts was a single piece of stone—a granite monolith 40 feet tall.

  - Roman citizens would have recognized that this characteristic gray granite was quarried in distant Egypt and transported at great expense across the Mediterranean to Rome. Thus, these columns displayed the vast power and wealth of the empire.

- Imperial power was also reflected in the triangular pediment. In Greek temples, the pediment was usually decorated with sculpture depicting scenes from mythology. But scholars speculate that, based on the pattern of holes for iron clamps, the sculpture in the Pantheon was probably an imperial eagle with laurel wreath—a characteristically Roman motif that, in this context, proclaimed unity between the imperial government and the gods.

- The dome of the Pantheon was the ancient world's most extraordinary structural achievement. Hemispherical in shape, it had a grid of four-sided indentations called **coffers** enlivening its inner surface and a 30-foot-diameter oculus at its apex. Within the cylindrical rotunda were two prominent cornices; a vaulted apse directly opposite the main entrance; six deep niches, each screened by a pair of marble columns; and forward-projecting shrines called ***aediculae*** for displaying statues of the gods.

© iStock/Thinkstock.

For the Romans, the great dome of the Pantheon would have evoked the heavens, covering the empire from horizon to horizon in unity with the gods.

- There was an elaborate pattern of colored marble veneer covering the wall and a simple square grid decorating the floor. Roman dominion was evident in the exotic marbles imported from conquered lands and even in that simple square floor grid, which suggests **centuriation**: the Roman practice of subdividing conquered territory into regular square parcels.

A Product of Engineering Genius

- Immediately below the Pantheon's pediment was a stone architrave bearing a prominent inscription: "Marcus Agrippa, son of Lucius, three times consul, built this." A Classical Greek temple would never have this sort of attribution; Greek temples were primarily a reflection of the polis, not the product of a powerful patron. Inscription notwithstanding, we are quite certain that Marcus Agrippa did not build the Pantheon, because the earliest brick stamps found in the structure date from A.D. 117—129 years after his death.

- Regardless of who designed the Pantheon, there is no doubt that Emperor Hadrian was its patron. There is considerable scholarly speculation that the designer of the Pantheon was actually Apollodorus of Damascus, Trajan's gifted engineer. Even if he was not, given the prominence of his prior engineering achievements, it is possible that his work strongly influenced the Pantheon's design.

- Circular buildings were common in antiquity. In one sense, the Pantheon represents a link in a long chain of development. It employed no revolutionary new technologies. Yet, in another sense, the Pantheon was innovative because it successfully met the structural challenges inherent in its extraordinary scale. The engineering of the Pantheon was unique in the ancient world. Indeed, not until the 19th century was a longer-spanning dome built successfully.

The Pantheon Dome

- The Pantheon dome was solid cast concrete, and with the exception of the stone columns and decorative elements, the remainder of the structure was entirely *opus testaceum.*

- A dome is a series of arches arranged around a central axis. The weight of the dome causes compression and lateral thrust. As radial segments spread laterally, gaps open up. Because concrete has very low tensile strength, a concrete dome tends to crack along these radial lines as a result of circumferential tension. Indeed, the Pantheon dome exhibits just this sort of cracking.

- The designer of the Pantheon responded to the challenge of lateral thrust with a variety of ingenious design features. Several of these reduced the thrust; others restrained it. The lateral thrust of a dome is directly proportional to its weight; thus, the obvious way to reduce thrust is by reducing the weight of the dome. Both the oculus and the coffers achieved this goal by simply removing material.

- The weight of the dome was also reduced significantly by modifying the components of the concrete itself. The Pantheon's dome used progressively lighter types of aggregate at progressively higher levels. There is even some evidence that sealed terra-cotta containers, or amphorae, were placed into the wet concrete to create air voids that further lightened the dome.

- But lightweight concrete would not have been strong enough to withstand the high concentration of internal forces around the oculus. For this reason, the oculus was reinforced with a 5-foot-thick ring of bricks. Just as we saw in the Domus Aurea, this ring functioned as a three-dimensional keystone, subjected to compression forces pressing inward from all sides. Its configuration was perfectly suited to this loading. With radially oriented bricks acting as voussoirs, the oculus ring carried load exactly like an arch—an ingenious arrangement that made this distinctive architectural feature possible.

The Cylindrical Drum

- Despite the devices to reduce weight, the dome still weighed more than 5,000 tons, generating immense lateral thrust. The most important mechanism for resisting the dome's thrust was the 20-foot thickness of the cylindrical drum walls supporting the dome. The drum was actually honeycombed with voids that effectively reduced its weight by half without compromising stability. Some of these voids—the entrance corridor, apse, and niches—were architectural features. The others were intended only to save concrete and provide access to the inner structure.

- The apse and niches presented a particular structural challenge. At each of these locations, the drum walls were reduced to one-third of their normal thickness and, thus, were significantly weakened. Yet overhead, the immense weight of the dome was applied uniformly around the perimeter of the drum. The weight transmitted down through the drum did not crush the weaker areas, however, because of a brilliantly conceived system of brick vaults and arches, embedded entirely within the drum wall.

- An internal skeleton of brick arches and vaults called **relieving arches** channeled the immense compressive forces around the niches into the adjacent wall sections and stone columns within the niches and, finally, down to the foundation—a solid ring of concrete 15 feet thick. This system distributed the building's immense weight safely to the underlying soil.

Veneer: A New Way of Using Stone

- Early Roman construction relied heavily on local tufa and travertine for traditional cut-stone masonry. As the new imperial architecture emerged, brick and concrete became the norm for load-carrying structure, while imported specialty materials, such as marble and granite, came into vogue almost exclusively for decoration.

- This trend arose in the 2nd century B.C., when Roman generals began commemorating their military victories by erecting monuments made of marble imported from conquered lands in the east. The fashion was facilitated by new technological developments—most notably, the invention of an iron saw for cutting hard stone into thin sheets.

- Emperor Augustus famously proclaimed that he had found Rome a city of brick and left it a city of marble. Over time, the acquisition and exploitation of decorative stone became a formal institution, with an extensive centrally controlled quarrying and transportation infrastructure providing a guaranteed source of supply for the emperors' grand projects.

- As this system matured, Roman quarries began prefabricating and stockpiling standard-sized columns and capitals. It was during this period that Romans acquired their taste for monolithic columns—each a single piece of stone turned on a huge lathe. This new style may have been influenced by a desire to show off the exotic multihued stones flooding in from the provinces, but it is also quite likely that monolithic columns became popular simply because they were so difficult to quarry, fabricate, and transport. In effect, they were public displays of Roman technological prowess.

The Pantheon Portico

- The Pantheon's front portico was massive, with its pediment and gabled roof supported on 16 monolithic columns of Egyptian granite. As impressive as these 40-foot, 60-ton shafts are, there is strong evidence that the portico was originally designed for even larger 50-foot columns. The evidence is in an odd gabled cornice on the front wall of the portico, 10 feet above the roof. Some scholars believe that this cornice corresponds to the originally planned level of the portico roof, which would then have required 50-foot columns.

- We do not know the reason for the change, but it is worth noting that a 50-foot column would weigh nearly twice as much as a 40-foot column—the 25 percent increase in both its length and diameter would nearly double the volume of stone used for the shaft. The substantial weight of the planned 50-foot columns may have caused insurmountable problems in quarrying, transporting, or erecting these shafts.

- Few buildings in the history of human endeavor have been as influential as the Pantheon. In 1632, when Pope Urban VIII called the Pantheon the "world's most celebrated edifice," he was probably referring not only to its fame but also to the host of buildings that were subsequently inspired by its architectural form.

Important Terms

aediculae: Covered niches, flanked by columns, intended as shelters for statues or shrines.

centuriation: The Roman practice of subdividing conquered territory into regular square parcels.

coffer: A polygonal indentation in a vault, dome, or ceiling.

portico: The covered front porch of a Classical-era building.

relieving arch: An arch built into a wall above a door or window opening to divert compressive force around the opening.

Suggested Reading

MacDonald, *The Architecture of the Roman Empire*, chapter 5.

———, *The Pantheon: Design, Meaning, and Progeny*.

Moore, "The Pantheon."

Taylor, *Roman Builders*, chapter 5.

White, *Greek and Roman Technology*, chapter 8.

Questions to Consider

1. How did the design of the Pantheon reflect the Roman Empire itself?

2. What were the principal structural innovations incorporated into the design of the Pantheon?

3. How does the Pantheon reflect the unique properties of concrete as a structural material?

Cities by Design—The Rise of Urban Planning

Lecture 10

In this second segment of lectures on Greek and Roman technology, we will study infrastructure development in the Classical world. The term "infrastructure" refers to large-scale technological systems that enhance societal functions, facilitate economic development, and enhance quality of life—particularly for urban populations. In classical antiquity, infrastructure systems included transportation networks, water supply systems, sewage systems, and facilities for public hygiene—most notably, the great bath complexes of imperial Rome. This lecture will focus on urban planning: the science of designing the overall layout, functional organization, and architectural character of a city or town.

Early Urban Planning … or Lack Thereof

- In Bronze Age Greece, small, independent urban communities typically centered on the king's palace and associated buildings, often clustered on a hilltop and surrounded by a wall for protection. Within this citadel, the streets were narrow, unpaved, and arranged in an irregular pattern dictated largely by the terrain. Outside the wall, the crude residences of common citizens were scattered about, with little organization or architectural character.

- Around the 7th century B.C., the polis became the center of political, cultural, and economic activity in the Greek world. Now, temples, public buildings, and markets took the place of the king's palace on the acropolis. As in the earlier citadels, though, the acropolis was served by an irregular pattern of streets and little evidence of a broader organizing scheme.

- The one major advance in **urban planning** during this period was the Greeks' deliberate efforts to enhance the architectural character of their public spaces. In a clear expression of civic pride, they placed elegantly designed temples, stoas, and council chambers in purposeful arrangements that demonstrated a keen sense of architectural balance, massing, and proportion.

Hippodamus of Miletus

- During the golden age of Athens, following the Greeks' great victory in the Persian Wars, urban planning took a monumental leap forward at the hands of one of the most interesting people in all of classical antiquity: Hippodamus of Miletus.

- Hippodamus was a philosopher and utopian theorist who attempted to define the perfect human community. He traveled throughout the Mediterranean, preaching the gospel of urban planning. Hippodamus promulgated his own well-conceived system based on a grid-like plan of rectangular city blocks formed by a network of major and minor streets and organized into well-defined public, private, and sacred zones. His ideas were so influential that, even today, Hippodamus is known as the father of urban planning.

- The meteoric rise of Athens in the 4th century B.C. as a naval and commercial power demanded the creation of a new port city. In 465 B.C., the Athenian assembly hired none other than Hippodamus of Miletus to plan the development of a port at Piraeus.

Port of Piraeus

- Hippodamus's plan for Piraeus was designed around a central axis extending along the ridgeline from Munychia Hill to the limestone quarries at the southern end of the peninsula. Perpendicular streets connected the harbors on opposite sides of the ridge. The grid of streets followed a strict hierarchy: 45-foot-wide main boulevards, 25-foot-wide secondary streets, and 15-foot-wide alleys.

- Sacred, public, private, and military zones were designated and marked with boundary stones. Within the public zone, Hippodamus placed the agora, council house, and public offices. In the sacred zone, he made accommodations for shrines and temples dedicated to the gods of foreign merchants who lived in the port.

- Development of the two naval harbors included hundreds of ship sheds, which protected the Athenian navy's famed triremes. In the private zone, uniformly sized residential blocks were subdivided into eight building lots, each accommodating a modest two-story dwelling with an enclosed courtyard. Hippodamus arranged these lots in stair-step fashion along the ridgeline so that every home had a view of the sea.

- For defense, the entire urban area was enclosed within a "great circuit": a thick stone defensive wall that was also integrated with the famous Long Walls, a pair of ramparts forming a protected corridor that connected Piraeus with Athens.

- Hippodamus's design for the Piraeus port was a masterpiece of order and functionality—a vivid manifestation of an urban planning revolution. The Greek world took note. From this point forward, new cities across the Mediterranean were nearly always designed according to the Hippodamus principles.

Port of Alexandria

- Further progress in urban planning occurred as formerly autonomous Hellenic city-states were subordinated to vast Hellenistic kingdoms. This progress was exemplified by the great port city of Alexandria—probably the grandest example of idealized urban planning in the ancient Mediterranean world. Developed at the end of the 4th century B.C. under the direction of a Greek architect named Dinocrates, Alexandria's city plan was greatly facilitated by its level site near the mouth of the Nile.

- The plan included a fully developed grid layout, with main boulevards 100 feet wide. It also incorporated a Jewish quarter—at the time, home to the ancient world's largest Jewish community. A mile-long artificial causeway connected the city to an offshore island, creating two fine sheltered harbors. On that island stood the Pharos of Alexandria—the famed lighthouse. Towering at 400 feet, it was one of the Seven Wonders of the Ancient World.

Romans Adopt Urban Planning

- The most coherent and comprehensive approach to urban planning was yet to come—in the Roman era. It is only natural that the Romans would excel in this endeavor: Their inherent love of order, organization, and standardization are characteristics intrinsic to effective urban planning.

- It is likely that the Romans first encountered urban planning during their conquest of the Etruscans, who were influenced by the Greeks and the Hippodamus city plan. The Romans adopted the grid system as their own but, consistent with their propensity to improve upon the ideas of others, then took urban planning to new heights—achieving a level of sophistication that would not be seen again until the Renaissance.

Aosta, a typical example of Roman urban planning, was founded in 25 B.C., when the legions of Marcus Terentius Varro crushed the Salassi tribe in northwestern Italy.

Aosta

- A typical example of the Roman planned city was Aosta in northwestern Italy. Unlike the Greeks, who established new cities on defensible hilltops, the Romans chose sites on flat, open ground astride strategically important roads or rivers. They compensated for the less-defensible terrain with better-designed defensive works.

- Aosta was surrounded by a stout rectangular circuit wall, augmented by towers at regular intervals. The urban area within was divided into four zones by two main streets—the ***cardo maximus*** (oriented north-south) and the ***decumanus maximus*** (oriented east-west). The four zones were then subdivided by a grid of streets into major residential blocks, called *insulae* (Latin for "islands"); each *insula* was then further subdivided into residential plots.

- Just off the intersection of the *cardo* and *decumanus* was the **forum**: the public marketplace. Other major public facilities—a theater, amphitheater, and baths—were all placed in one quarter of the town. A well-developed system of water supply and drainage was integrated within and beneath the grid of streets.

- The city plan of Aosta was coherent, comprehensive, and highly functional. Yet perhaps its most important characteristic was that it was entirely typical. Scores of Roman cities throughout the empire—in Italy, Spain, Gaul, North Africa, and the Middle East—followed the same basic model, with only minor variations to account for differing terrain, climate, and population.

The Roman Military Camp

- The Roman city planning model was also applied in a totally different context—the Roman military camp, or ***castrum***. The Roman legions built these camps as operating bases while campaigning on the frontiers of the empire. *Castra* were highly standardized—perhaps even more so than Roman cities.

- The typical *castrum* was essentially identical to the Roman planned town, except that a large headquarters building called the ***principia*** was placed at the center, and the *insulae* were replaced by barracks for the troops and stalls for the horses. Many *castra* had hospitals, granaries, and even bath complexes.

Land Surveying

- Essential to effective Roman urban planning was the system of land surveying—a system that clearly demonstrated the Roman propensity for standardization and organization. Surveying involves three operations: establishing a **bearing**; measuring distance along that line; and measuring the elevation of a point on the ground with respect to a reference point.

- To establish a bearing, the Roman surveyor—***agrimensore***, or "land measurer"—used an ingenious device called a ***groma***. The device consisted of a pivoting cross-staff mounted on an L-shaped bracket, which was itself mounted on a post with a metal shoe at the bottom. At the ends and center of the cross-staff were weighted strings, called plumb lines.

- The *agrimensore* used a variety of tools for measuring distances. For relatively short distances, he would have used a ***decempeda***—a hardwood measuring rod 10 Roman feet long, capped with special bronze fittings that allowed multiple rods to be connected end to end.

- For longer distances, surveyors used ropes coated with wax or pitch to minimize shrinkage. These ropes would have been fabricated in standard lengths—most commonly the ***actus***, equal to 120 Roman feet.

- For extremely long distances—such as those associated with intercity roads and aqueducts—the *agrimensore* used an ingenious device called a ***hodometer***. This machine was a wheeled cart, with its axle driving a set of cog-wheels, calibrated so that a small pebble would drop into a bucket after the cart had rolled exactly 1 Roman mile—a distance of 5,000 Roman feet.

Centuriation

- Another fascinating demonstration of Roman surveying prowess is centuriation—the quintessentially Roman practice of subdividing conquered territories into square parcels of farmland for allocation to settlers. The centuriation process began with the establishment of a new intercity road—which was often built by army engineers as the Roman legions gained control of new territory. With this road as a baseline, the surrounding countryside was divided into large square parcels, called *centuriae*, or centuries.

- The ***centuria*** was so named because it was divided into 100 smaller square plots, called ***heredia***. Each of these was further divided in half along the north-south axis to create a rectangular plot, the

jugerum, defined as the area of land that could be plowed by one pair of oxen in one day. This system was so effective that it remains in use today as the basis for defining property lines in many regions of rural Italy.

Important Terms

actus: A Roman unit of land measurement, equal to 120 Roman feet.

agrimensore: A Roman surveyor ("land measurer").

bearing: In surveying, a line oriented in a particular direction and passing through a designated reference point.

cardo maximus: The main north-south street in a Roman town or city.

castrum: A Roman military camp.

centuria: In Roman surveying, a square plot of land measuring 20 *actus* by 20 *actus*.

decempeda: A Roman measuring rod, made of hardwood and capped with bronze fittings that allowed multiple rods to be connected end to end.

decumanus maximus: The main east-west street in a Roman town or city.

forum: The central public marketplace in a Roman town or city.

groma: A Roman surveying instrument, used for laying out a rectangular grid.

heredia: In Roman surveying, a square plot of land measuring 2 *actus* by 2 *actus*.

hodometer: A cart-mounted device used to measure long distances over land.

jugerum: In Roman surveying, a rectangular plot defined as the area of land that could be plowed by one pair of oxen in one day.

principia: The headquarters building at the center of a Roman *castrum*.

urban planning: The science of designing the overall layout, functional organization, and architectural character of a city or town.

Suggested Reading

Adam, *Roman Building*, chapter 1.

Hale, *Lords of the Sea*, chapter 8.

Oleson, *The Oxford Handbook of Engineering and Technology*, chapter 25.

Questions to Consider

1. How did the Greeks and Romans differ in their approach to urban planning?

2. How are Greek and Roman ideas about urban planning still evident in modern cities?

3. What are the relationships between the Romans' propensity for large-scale military conquest and their proficiency in urban planning and surveying?

Connecting the Empire—Roads and Bridges

Lecture 11

In the late 1st century B.C., the Greek historian Dionysius of Halicarnassus wrote, "The extraordinary greatness of the Roman Empire manifests itself above all in three things: the aqueducts, the paved roads, and the construction of drains." At its height, the Roman Empire had 75,000 miles of public roads, organized into a system that incorporated nearly 1,000 bridges (many of which still survive). Although this incredible system was intended primarily to project military power, it also yielded second-order benefits: economic development, enhanced public order, and dissemination of Roman culture and law. In the ancient world, infrastructure mattered, and Roman engineers were seen as particularly adept at designing and building it.

Historical Development of Roads

- In antiquity, an intercity road was both a political and a technological entity. In fact, the principal purpose of many ancient roads was to strengthen political control over a region. Constructing a road also required a substantial commitment of state resources, which required considerable political authority. Before the Roman era, the most significant technological developments in land transport occurred under the patronage of powerful kings.

- Around 500 B.C., King Darius of Persia developed an extensive permanent road system—an instrument for exercising control over his vast empire. This road system included the famous Royal Road, which was the basis for a sophisticated courier system. Yet despite its superb organization, the Royal Road reflected relatively primitive technology. It was unpaved, and many streams and rivers along the route could be crossed only at fords or on ferryboats.

- Emperor Augustus capitalized on the strategic value of Roman roads by creating an official courier system, the ***cursus publicus***, supported by a network of roadside facilities for the couriers and their horses.

- The Classical Greek city-states achieved prosperity and influence largely through maritime commerce and naval power. Greece's rough terrain and the remote location of many of the city-states made land transport inherently difficult. Greek roads were usually little more than dirt paths, suitable for pack animals and pedestrians but not for vehicles. The only major exceptions were the sacred roads that connected major religious sanctuaries with their host cities—for example, the Sacred Way, a well-built road from Athens to the sanctuary at Eleusis.

The Via Appia

- In sharp contrast to Greece, the Roman Republic established a strong tradition of road building, primarily to support military operations, first in Italy and then throughout the Mediterranean world. The first major Roman road was the Via Appia (or Appian Way), built in 312 B.C.

- Establishing a precedent that would become a characteristic feature of nearly all Roman intercity roads, the Via Appia ran absolutely straight for long stretches, entirely independent of the topography over which it was constructed. What's more, the Via Appia bypassed most of the towns along its route, connecting to them only with branch roads. In these aspects of its design, the Via Appia was similar to a modern interstate highway.

- In one sense, the Romans' use of these relentlessly straight roads simply reflects the fact that the shortest distance between two points is a straight line. But it also confirms that these roads were intended primarily to facilitate the rapid movement of infantry. Building a straight road across hilly country would inevitably produce some steep **gradients**. Infantry would not be appreciably impeded by these gradients, but horse- or ox-drawn vehicles certainly would.

- The straight road also became a statement of Roman dominion—over both geographic regions and nature itself. This powerful statement even had its own unique punctuation marks: carved milestones placed at regular intervals along the route and triumphal arches marking the beginning and end of each road, the entrances to major cities, and even some bridges.

A Multifaceted Engineering Challenge

- Designing and building these roads was a multifaceted engineering challenge. Selecting the route required a careful topographic and geologic analysis of the land over which the road would pass. Once the route was chosen, surveying and marking the route presented an even greater challenge—because those long, straight stretches of road rarely provided a direct line of sight between their start and end points.

- The chosen route would then dictate the need for a host of associated engineering works, many of which needed to be initiated before construction of the roadway proper could begin. These included bridges and viaducts to cross waterways and low-lying areas, drainage structures to divert small streams, land reclamation projects to eliminate swamps and marshes, rock-cut terraces and retaining walls along valleys and mountainsides, and tunnels through mountains or ridgelines.

Construction of the Roadway

- After the supplemental structures were fully worked out, construction of the roadway began, using the following process:
  - Dig two narrow trenches.
  - Set upright stones to form curbs, which were 14 Roman feet apart, wide enough for two carriages.
  - Excavate between curbstones to a layer of stable soil or rock.
  - Put down a layer of large stones for strength and drainage.
  - Put down a layer of gravel mixed with clay for enhanced waterproofing.
  - Create a camber (curve) to improve drainage.

The Roman tradition of road building was originally established to support military operations—primarily, the rapid movement of infantry.

- Cap with a "wearing surface" of compacted pebbles. (Later, most important roads were resurfaced with a pavement of polygonal basalt blocks, wedged into place.)
- Add smaller stone blocks to delineate 10-foot shoulders.
- Surface with compacted gravel.

- Although this structure was typical, there was really no such thing as a "standard" Roman road. Configurations varied tremendously—from road to road and even from location to location along the same road. The variations reflected deliberate adaptations to different soil types, moisture conditions, and availability of construction materials.

Building Bridges

- Of all the factors contributing to the efficiency and majesty of the Roman road system, none was more important than Roman engineers' mastery of bridge design and construction. Previous civilizations had failed to develop this competency; thus, as we saw with the Persian Royal Road, the effectiveness of their land transport was fundamentally compromised by the need to cross rivers by ferry or ford.

- The Romans succeeded where others had failed by exploiting a technology we already know well: the arch, which had sparked a revolution in Roman architecture. In the hands of bridge engineers, the arch allowed those arrow-straight Roman roads to defy the terrain. The first great Roman bridges were built in the 2nd century B.C., when the arch was just coming into widespread use in Roman construction.

Ponte Milvio

- An excellent early example of Roman bridge building was the Ponte Milvio, built around 110 B.C. to carry the Via Flaminia over the Tiber River just north of Rome. The Ponte Milvio had four main arches, each spanning about 60 feet and constructed entirely of cut stone—durable travertine on the outer faces and softer tufa within. These arches were supported on massive prow-shaped piers, which channeled the waters of the Tiber around the foundations. Above the piers were smaller arched openings, which relieved pressure on the sides of the bridge during flooding.

- The greatest challenge in building a bridge like the Ponte Milvio was constructing the stone piers in the riverbed. Fortunately, Vitruvius provides us with a clear description of how this was done. Today, this same basic process, with only minor modifications, is still used to build underwater bridge piers.
  - First, a structure called a **cofferdam** was fabricated from sharpened oak posts lashed together and then driven into the riverbed at the pier location. A slightly smaller cofferdam was installed inside the first, and the space between the two walls

was packed full of clay to create a waterproof barrier. The river water was then pumped out of the inner cofferdam.

  - With the riverbed exposed, workers inside the cofferdam could excavate down to a firm layer of rock or soil. If necessary, piles made of rot-resistant wood were driven into the soil to provide additional support, and then the pier was built—usually of cut stone with rubble-and-mortar fill. Once the top course of stone exceeded the water level, the cofferdam was removed and the remainder of the structure was completed.

Use of *Opus Testaceum* and Concrete

- By the first half of the 2nd century A.D., the Roman construction revolution had taken hold, a bold new paradigm in building design had emerged, and bridge design was evolving in the same direction—toward the total replacement of stone structural elements with *opus testaceum* and concrete.

- A typical example was a viaduct carrying the Via Appia across a broad valley near the town of Sessa Aurunca. This structure was more than 500 feet long and consisted of 21 arches made entirely of brick and concrete. Just as Trajan's Market set a new standard for functionality and construction efficiency in building design, so the viaduct at Sessa Aurunca reflected the application of this new state of the art to bridges.

A Temporary Bridge of Permanent Significance

- Although most of the bridges associated with the Roman road system were permanent structures made of stone or brick, a temporary timber structure of great historical importance is worth mentioning: Julius Caesar's bridge over the Rhine River, built in 53 B.C. during his celebrated Gallic campaign.

- The structure was supported on a series of frames (called **trestles**) created by using a barge-mounted pile driver to hammer sharpened timbers into the riverbed. The piles were connected in pairs and driven in at an angle to resist the current. A massive 2-foot-wide timber beam was held in place with cross-members. There was an additional angled pile on the downstream side, lashed in place, and a cluster of piles upstream to deflect logs or boats that might destroy the bridge. Finally, Caesar's engineers installed bundles of wooden sticks across the beams to create a deck on which the troops could march.

- Caesar's bridge over the Rhine was indisputably an engineering marvel, yet one could argue that a short-term structure built to support a single military operation does not belong in a lecture on civil infrastructure. But to the Romans, roads and bridges were more than just civil infrastructure. These structures were statements of Roman power: proclamations of dominion over a territory and its people. In that sense, Caesar's bridge is just another chapter of the same story—albeit a particularly dramatic one.

Names to Know

cofferdam: A temporary structure used to construct bridge piers and port facilities underwater.

cursus publicus: The official Roman state courier system, established by Emperor Augustus.

gradient: The slope of a road or aqueduct channel, expressed as a percentage: elevation change per horizontal distance.

trestle: A wooden frame that supports the deck of a timber bridge structure.

Suggested Reading

Oleson, *The Oxford Handbook of Engineering and Technology*, chapter 22.

Staccioli, *The Roads of the Romans*.

Ulrich, *Roman Woodworking*, chapter 5.

White, *Greek and Roman Technology*, chapter 8.

Questions to Consider

1. What are the relationships between the Romans' propensity for large-scale military conquest and their proficiency in road and bridge construction?

2. What sorts of engineering work would be required to support a Roman road construction project, beyond the road itself?

3. How did the design and construction of Roman bridges mirror that of Roman buildings through the republican and imperial eras?

From Source to City—Water Supply Systems

Lecture 12

Of course, water is essential for human life, but it's not just for drinking. We use it to irrigate our crops and to generate power; we use it for transportation, sanitation, food processing, recreation, firefighting, cooling, and a wide variety of industrial purposes. And though these are modern uses of water, they were all prevalent in antiquity, as well. Water supply systems played a crucial role in ancient technological development. In this lecture, we'll look at the historical development of water supply technologies.

The Hydrologic Cycle

- Water supply is a human endeavor, but it occurs within the context of a natural process called the **hydrologic cycle**. Freshwater falls to the earth as precipitation. After reaching the ground, rainwater takes three different forms:
    - **Surface water** flows over the earth's surface and is channeled into progressively larger streams and rivers; it may be temporarily stored in lakes and oceans.
    - **Soil water** infiltrates into the earth but is retained near the surface, in the voids between soil particles.
    - **Groundwater** infiltrates more deeply into the soil and is ultimately collected in an **aquifer**—a porous stratum of soil or rock fully saturated with water. An aquifer is generally formed immediately above an impermeable stratum, called an **aquiclude**, which prevents the water from percolating any farther downward. The upper surface of an aquifer is called the **water table**.
- If the stratum intercepts the surface, it forms a spring. Springs were the principal sources for most major water supply systems in the ancient world.

Early Water Supply Technologies

- Ancient cities were usually located on or near rivers; thus, one might expect that these rivers would serve as the principal water sources. But because rivers flowing through ancient cities were invariably used for disposal of human, animal, and industrial waste, they were generally too polluted to be used for drinking water. Also, because a river typically flows through a city at a very low elevation, large quantities of water would have to be lifted—a significant technological challenge.

- By far, the most common water supply technology in the ancient world was the well. A well is a vertical shaft dug down into an aquifer, then stabilized with an inner wall of wood or masonry. Because the aquifer is saturated, water collects at the bottom of the well shaft, up to the level of the water table, and can be drawn out with a rope tied to a bucket. Primitive wells from the 9th and 8th millennia B.C. have been found on the island of Cyprus—and wells have been in continuous use since then.

- Over the centuries, the need to lift water more efficiently stimulated a number of important technological developments. The **shadoof**, a counterweighted beam pivoting on a vertical post, has been used for lifting water since at least 2000 B.C.

- Where there was no accessible aquifer, the **cistern** was the preferred alternative to a well. A cistern is a masonry tank, usually located just below ground level and used to collect rainwater from a roof or paved surface.

Early Public Water Supply Systems

- Some of the world's first large-scale public water supply systems were built by the Assyrians. Around 700 B.C., King Sennacherib built a dam to divert the Atrush River into a manufactured channel that carried water to the city of Nineveh more than 30 miles away. This stone channel is often regarded as the world's first true **aqueduct**.

- Around this same time, Persians living on the arid Iranian plateau developed the **qanat**—one of the most important advances in the history of water engineering. A qanat is a tunnel driven into a hillside to tap an underground aquifer. It consisted of a mother well and a horizontal tunnel dug from the base of the hill toward the mother well, on a slight uphill gradient. Along the way, a series of vertical shafts were added at regular intervals. Some qanats were more than 20 miles long and had as many as 700 vertical shafts.

- The cities of Archaic- and Classical-period Greece generally did not build large-scale public water systems. The typical Greek polis grew up around a small spring; as the population of the polis grew, the spring water was supplemented with water piped in from sources outside the city walls. Greek pipelines were usually made of terra-cotta segments, 8 to 10 inches in diameter, fitted together end to end, sealed with plaster, and set in a trench.

The aqueduct built by Eupalinos was described by the historian Herodotus, who regarded it as one of the ancient world's greatest engineering achievements.

- A few city-states experimented with an innovative alternative to tapping water sources outside their walls. They built storage reservoirs consisting of a stone enclosure and a flat roof. Water from a spring entered on one end; at the opposite end, it flowed into a trough, where people could fill their **amphorae**.

- Although Hellenic-period water systems generally did not exhibit technological sophistication beyond the simple storage reservoir, there was one major exception to this rule. In the 6th century B.C., on the island of Samos, a Greek engineer named Eupalinos built an aqueduct that included an extraordinary 3,400-foot tunnel through the solid rock of Mount Ampelos—an astonishing engineering achievement.

- The development of water supply systems advanced considerably during the Hellenistic era. Technological developments were aided by Hellenistic science, as the flourishing Museum at Alexandria provided important new insights about hydraulics and mechanics during this period.

The Inverted Siphon

- Hellenistic aqueducts were substantially larger and more sophisticated than their Hellenic forebears; indeed, they were quite comparable to later Roman systems in all respects but one. Because Hellenistic engineers failed to exploit the arch as a structural element, they were not able to raise a water channel significantly above ground level. Thus, when crossing a deep valley, their only practical option was to use an **inverted siphon**.

- An inverted siphon consists of pipeline extending between two tanks—header and receiving—both built of stone, lined with a coating of waterproof mortar. In a conventional **open-channel aqueduct**, when a spring gushes forth, water flows downhill from the source to the settling tank to the **header tank**. In an inverted siphon system, when the spring gushes forth, it seems to be flowing uphill.

- The reason is that the two tanks and pipeline constitute a closed system, and within this system, the water is flowing downhill—from the header to the **receiving tank**. Because the pipeline starts and ends submerged within a reservoir, it is flowing full and under pressure. This pressure is caused by the difference in the water levels of the two tanks. Water will continue to flow through the pipeline under pressure as long as the water level in the header is higher than the water level in the receiving tank.

- Archaeologists have discovered at least 20 of these inverted siphon aqueducts in the Hellenistic east. The most spectacular, by far, is the one at Pergamon. The size of this system was absolutely unprecedented in antiquity, as was its configuration. It supplied more than a million gallons of water per day from a spring 25 miles north of the city. The final stretch of the system was an extraordinary inverted siphon crossing a valley 2 miles long and 650 feet deep.

Lead Piping

- The pipeline at Pergamon represents a successful solution to an unprecedented engineering challenge. Pipe pressure increases in proportion to the vertical distance between the pipe and the **hydraulic gradient**—an imaginary line connecting the water surfaces of the two reservoirs. Based on the topography of the valley at Pergamon, we know that the maximum pressure in the siphon was about 250 psi (pounds per square inch)—substantially higher than in any other ancient pipeline.

- Segmental stone pipes would not have worked for the unique elevated configuration of the Pergamon siphon. Based on the spacing and shape of its stone posts, it is clear that its pipe was made of metal. For much of the 20th century, scholars believed the metal to be bronze, because the only alternative—lead—was not strong enough for a high-pressure pipe. Then, in 1976, a chemical analysis of the soil along the path of the siphon indicated a lead concentration 56 times higher than normal—proving that the pipe was, in fact, made of lead.

The Romans' Practical Approach

- The sophistication of the inverted siphon system was constrained by the Hellenistic engineer's failure to exploit the arch as a means of elevating a water channel. Roman engineers knew no such constraint. They carried an aqueduct channel across a valley in a very different way—using an arched bridge.

- The Romans preferred bridges over siphons, not for lack of knowledge, but for a quintessentially Roman reason: practicality.
  - For all its theoretical superiority and technological sophistication, the ancient inverted siphon must have been a maintenance nightmare. The siphon at Pergamon had several thousand joints between pipe segments; when the system was pressurized, every one of these joints was a potential leak.

  - And because the inverted siphon is a closed system, it would have been particularly difficult to inspect and maintain. This problem was greatly exacerbated by the high mineral content of many ancient water sources.

- Limestone strata occur quite frequently in the geology of the Mediterranean region. When rainwater percolates down through limestone to become groundwater, calcium ions leach from the limestone into the water. As this water moves through an aqueduct, these minerals precipitate out as **calcium carbonate**, a white solid that accumulates on the inside surfaces of the conduit.

- Roman open-channel systems were not immune to this problem—but they were purposefully configured so that a maintenance worker could physically enter the channel to remove mineral deposits with a hammer and chisel. This was not possible with the closed pipes used in an inverted siphon; thus, these systems were likely to clog after a few decades of use and require total replacement. The Romans' apparently extravagant arcaded aqueduct bridge was actually the more pragmatic way to carry a water channel across a valley.

- In the next lecture, we'll gain a fuller appreciation for the Romans' beautifully integrated water infrastructure system by examining some of Rome's great aqueducts and experiencing the process of aqueduct design.

Important Terms

amphora: A terra-cotta jar, used to transport and store liquids.

aquaclude: An impermeable stratum of soil or rock, over which an aquifer forms.

aqueduct: A manufactured structure that carries water from a distant source to a city or town.

aquifer: A geologic formation consisting of a porous stratum of soil or rock that is fully saturated with water.

calcium carbonate: A solid white substance that accumulates on the inside surfaces of pipes carrying water with high mineral content.

cistern: A masonry tank, usually located just below ground level and used to collect rainwater from a roof or paved surface.

groundwater: Water that infiltrates deeply into the soil and is ultimately collected in an aquifer.

header tank: The water reservoir at the start of an inverted siphon.

hydraulic gradient: An imaginary line connecting the water surfaces of the header and receiving tanks in an inverted siphon system; used to calculate pressure in the pipeline.

hydrologic cycle: The natural process by which water falls to the earth in the form of precipitation; flows over and into the soil; is transported by streams and rivers; is stored in lakes and oceans; and ultimately, returns to the air by evaporation and transpiration.

inverted siphon: A type of aqueduct (or segment of an aqueduct) used to transport water across a valley through a pipeline under pressure.

open-channel aqueduct: An aqueduct in which the water flows on a continuous downhill gradient and does not flow under pressure.

qanat: A tunnel driven into a hillside to tap an underground aquifer.

receiving tank: The water reservoir at the end of an inverted siphon.

shadoof: A primitive water-lifting device, consisting of a counterweighted beam pivoting on a vertical post.

soil water: Water that infiltrates into the earth but is retained near the surface, in the voids between soil particles.

surface water: Water that flows over the earth's surface and in rivers and streams.

water table: The upper surface of an aquifer.

Suggested Reading

Hodge, *Roman Aqueducts and Water Supply*, chapters 1–8.

Landels, *Engineering in the Ancient World*, chapter 2.

Oleson, *The Oxford Handbook of Engineering and Technology*, chapter 11.

Questions to Consider

1. Why were Classical Greek cities able to thrive without the large-scale water supply systems that would characterize later Roman cities?

2. What technological limitation prevented Hellenistic water supply systems from achieving the same scope and sophistication as Roman systems?

3. Why did Roman engineers shun technologically sophisticated inverted siphon systems in favor of simpler but more expensive aqueduct bridges?

Engineering a Roman Aqueduct

Lecture 13

Here's how Roman engineers created a typical aqueduct: use a spring as the water source; employ the continuous downhill flow of water through an artificial channel; build an arched bridge to carry the channel across a valley; deliver the water into a receiving tank within the city; distribute the water to fountains and baths through a pipe network; and finally, discharge the wastewater into a local stream. Although this model illustrates the basic functions of a Roman aqueduct, it does not adequately capture the vast scope and technological complexity of this amazing infrastructure system. In this lecture, to deepen our understanding of Roman aqueducts and to better appreciate the challenges faced by their engineers, we'll explore the design process itself.

Determining the Water Source and the Best Route

- Roman engineers charged with developing the design for a new aqueduct began the process by identifying the location where the aqueduct would deliver water to the city. Within the city, water distribution must be driven by gravity; thus, the delivery point should be at or near the city's highest point, and the channels must maintain a downhill gradient all along their length.

- The gradient of a channel is normally expressed as a percentage—elevation change per horizontal distance. For example, the average gradient of the aqueducts supplying Rome was about 0.4 percent, meaning that the elevation of the channel decreases by 0.4 feet (about 5 inches) for every 100 feet of length. Because of the need to accommodate varying terrain, most aqueducts did not use a single uniform gradient; rather, they were designed in several segments.

- The next task was to locate an appropriate water source. A few Roman aqueducts drew surface water from rivers, but most tapped groundwater from upland springs. After finding a possible water source, workers dug a shaft downward until locating the water table, then drove a horizontal gallery inward from a lower point on the hillside, allowing water to flow freely from the aquifer to the surface. At the point where this gallery emerged, they built a settling tank: a masonry basin designed to remove sediment from the water before it entered the aqueduct.

- In order to determine the best route for the aqueduct, Roman engineers made a careful survey of the terrain along all possible routes. They performed this survey with a device called the ***chorobates***—a long wooden table used to measure the change in elevation over a distance. Looking though a pair of sights toward a target rod held by an assistant some distance away, surveyors were able to measure the height of the sight line on the rod.

- By repeating this measurement hundreds of times along the proposed route, engineers developed a **terrain profile**, or a graph showing the variation in elevation along the route. The terrain profile was used to evaluate the suitability of the proposed aqueduct channel alignment.

The Aqueduct Channel

- A typical aqueduct channel was a rectangular concrete trough, roughly 3 feet wide and 5 feet high, covered with a stone vault. The walls and floor were coated with a special mortar made of lime and crushed brick and polished to a glasslike finish—to reduce friction and to facilitate the removal of calcium carbonate deposits. The vaulted roof reduced evaporation and prevented foreign material from falling into the channel.

- Wherever possible, the channels were built slightly below ground level, using a technique called **cut and cover**. Using this method, the channel was constructed inside a trench, with inspection shafts added at regular intervals, and then the trench was backfilled with soil.

- Even though this below-ground configuration was the norm, it was sometimes necessary to raise the channel above ground to maintain an appropriate gradient. Where the channel needed to be raised 6 feet or less, engineers used a ***substructio***—a solid stone- or brick-faced embankment with the water channel perched on top. Where the required channel elevation was greater than 6 feet, engineers used either a bridge or an inverted siphon.

- Aqueduct bridges were used for two distinctly different purposes. The first was to carry the water channel across a valley. The second was to elevate the channel over relatively level ground—often for long distances—simply to maintain the planned gradient. Single-tier arcades were generally used whenever the required channel elevation was less than 70 feet. For greater heights, two-tier or three-tier arcades were used for improved stability.

- Had the Romans used single-tier arcades for aqueducts over 70 feet tall, the piers would have been quite slender and therefore dangerously susceptible to **buckling**. By adding arches at multiple levels, each long pier was subdivided into three shorter ones—and the load-carrying capacity of the structure was substantially improved.

The Inverted Siphon

- Based on archeological evidence, it appears that no Roman aqueduct bridge ever exceeded 160 feet tall. On the other hand, Roman engineers routinely built inverted siphons across valleys deeper than 160 feet. Siphons were used almost exclusively in situations where bridges would have been impractical.

- The most spectacular Roman inverted siphon is found on the Gier aqueduct in Lugdunum (Lyon), France. Although somewhat smaller than the Hellenistic siphon at Pergamon, the Gier aqueduct's capacity was six times greater—an astonishing 6.6 million gallons per day.
    - At the top of the valley, water from a conventional open-channel aqueduct poured into a vaulted header tank and then into nine parallel lead pipes, each 10 inches in diameter. These pipes ran down a massive concrete ramp, then descended into the valley underground.

The famous Pont du Gard in southern France is 160 feet tall, which seems to have been the upper limit for Roman aqueduct bridges.

- Near the bottom of the valley, the pipelines emerged from underground; crossed the valley on a low arcaded structure called a venter bridge; and climbed the opposite side to a receiving tank, where another conventional open channel continued on to Lugdunum. The purpose of the venter bridge was to reduce pressure in the siphon pipes by raising them above the valley floor.

Building Tunnels

- The Romans built tunnels only when all other alternatives proved impractical. According to Vitruvius, the standard Roman tunneling technique was to excavate a horizontal gallery in conjunction with a series of vertical shafts.

- The vertical shafts simplified surveying by allowing the alignment of the tunnel to be controlled precisely from the surface. The vertical shafts also reduced construction time by allowing for multiple work spaces within the tunnel and by providing multiple pathways for removing excavated material.

Rome's Major Aqueducts

- Rome's first major aqueduct was the Aqua Appia, commissioned in 312 B.C. by Appius Claudius, the magistrate who also built the Via Appia, Rome's first major intercity road. The Aqua Appia provided a much-needed supplement to the city's wells, cisterns, and springs, which were becoming increasingly inadequate for a rapidly growing population. Its channel ran entirely underground, from a spring 10 miles east of Rome to a fountain in the Forum Boarium—one of the city's lowest public spaces.

- About 40 years later, the Anio Vetus was added to the system. This aqueduct drew water from the Anio River in the mountains 40 miles east of Rome. It, too, ran underground, but by following higher terrain, it was able to deliver water to the Viminal Hill on the northeastern side of the city.

- In 145 B.C., the longest and most celebrated of all Roman aqueducts—the Aqua Marcia—was completed. It ran underground for 50 miles from its source, a spring in the Anio Valley. But as it left the mountains, the channel was elevated on an arcade for the 6-mile stretch across the plain of Campagna into Rome. This first-ever use of an aqueduct bridge provided the elevation needed to deliver water to the Capitoline Hill.

- As demand continued to increase, the Aqua Tepula was added in 127 B.C. The Aqua Julia, named for its patron Julius Caesar, followed in 33 B.C. These two aqueducts originated at separate sources in the Alban Hills south of Rome, but when they reached the Campagna plain, they were both piggybacked on top of the Aqua Marcia, creating a triple-decker bridge. The Aqua Claudia and Anio Novus, both built in A.D. 52, used this same expedient.

- Even though aqueduct channels ran adjacent to each other for many miles, the Romans generally did not combine the water from multiple channels into one.
    - Merging channels might have made sense from an engineering perspective, but the Romans kept them separate—primarily as

a matter of taste. Each aqueduct had its own source, and the water from each source had its own special character.

  - The Aqua Marcia was renowned for supplying Rome's best-tasting water. The Anio Novus supplied sediment-filled, foul-tasting river water that was considered to be suitable only for irrigation.

- By the 3rd century A.D., Rome was served by an integrated system of 11 aqueducts sprawling over 300 miles and delivering about 300 million gallons per day. By comparison, the New York City water supply system would not achieve this capacity until well into the 20th century. When we consider both the effectiveness and the majesty of these aqueducts, it is hard not to agree with the historian Edward Gibbon, who called them "the noblest monuments of Roman genius and power."

Important Terms

buckling: Stability failure of a structural element subjected to compression.

chorobates: A Roman surveying device used to measure the change in elevation over a distance.

cut and cover: Method used to construct an underground aqueduct channel just below ground level.

substructio: A stone- or brick-faced embankment used to support an aqueduct channel a few feet above ground level.

terrain profile: A graph showing the variation in elevation along a route.

Suggested Reading

Hodge, *Roman Aqueducts and Water Supply*, chapters 4–8.

Landels, *Engineering in the Ancient World*, chapter 2.

Oleson, *The Oxford Handbook of Engineering and Technology*, chapter 11.

Questions to Consider

1. What factors would a Roman engineer have considered in selecting a water source for a new aqueduct system?

2. Under what circumstances would it make sense to design an aqueduct channel with a relatively high gradient? A relatively low gradient?

3. What factors would a Roman engineer have considered in deciding whether an aqueduct channel should follow the contours of the terrain, be elevated on a *substructio* or bridge, or employ a tunnel or inverted siphon?

Go with the Flow—Urban Water Distribution

Lecture 14

This lecture deals with the Romans' sophisticated system of water distribution and sewage disposal. The best way to study it is simply to "go with the flow" and follow the water from the aqueduct all the way to the sewer that carried it out of the city. Once the Roman aqueducts reached their destination, the water was cleaned, then distributed to baths, fountains, industrial consumers, and private users all over the sprawling city. The supply network needed mechanisms for measuring usage so that private customers could be appropriately charged for their water. And ultimately, to keep the city from drowning in its own effluent, a parallel system for collecting and disposing wastewater was essential.

De Aquaeductu

- Much of our information about the Roman water supply system comes from Sextus Julius Frontinus, a well-respected member of the senatorial class. He held the Roman consulship three times before the emperor Nerva appointed him water commissioner of Rome in A.D. 97. Upon assuming his duties as commissioner, Frontinus devoted himself to learning everything he could about Rome's water system. And then he did something extraordinary: He wrote it all down in an official report called *De Aquaeductu*, which has survived to the present day.

- As Frontinus makes clear, the most important characteristic of the Roman water supply—and its most important difference from modern systems—is continuous flow. Except for occasional maintenance and repair, an aqueduct was never shut off. Water moved through the urban distribution system without interruption—usually with no long-term storage along the way.

The Settling Tank

- At or near its delivery point, an aqueduct typically emptied into a **settling tank**, which was designed to remove sand, silt, and other suspended solids from the water before it entered the urban distribution network. A settling tank was often built at the aqueduct's source, as well.

- A settling tank operated according to the principle that moving water can carry more suspended solids than still water. Water in the aqueduct channel easily carried sediment; in the chambers, it slowed almost to a halt, and sediment settled to the bottom, where it was periodically cleaned out. At the top of a second chamber was an outlet that drew water off at the same rate as it entered the first chamber—maintaining continuous flow.

The *Castellum Divisorium*

- After passing through the settling tank, the water was directed into the ***castellum divisorium***—a structure designed to divide the incoming flow into multiple channels. In an example that has survived at Pompeii, water from the aqueduct flowed into a shallow circular basin about 18 feet in diameter. The flow first passed through two metal gratings—one coarse, one fine—to remove debris. Then, it was divided into three channels, leading to three outlets: large lead pipes that served as Pompeii's water mains.

- The water attendant used three movable wooden gates to control the flow into each channel. Scholars have suggested that the three gates were of different heights, creating a built-in priority system for water distribution. According to this theory, the lowest gate controlled the water supply to the public fountains; the next-higher gate provided a lesser supply to the public baths; and the highest gate allocated the smallest quantity of water to private homes.

- The pipes from the *castellum divisorium* conducted the water to a series of **secondary *castella***—each consisting of a brick pier topped by a 250-gallon tank made of lead. The open-topped tank was continuously filled by a large pipe running up the side of the tower and continuously emptied by smaller pipes running back down to public fountains and private homes in the immediate vicinity.

- The principal purpose of the secondary *castellum* was to control pressure in the distribution system. The system was essentially a series of inverted siphons, with each secondary *castellum* serving simultaneously as both the receiving tank for the siphon pipe above and the header tank for the one below. Each tank was lower than the one that fed it; thus, the entire system was propelled by gravity.

Lead Piping

- At the heart of this system were the pipes themselves. In the Roman world, water supply pipes could be terra-cotta, stone, or even wood, but most often, they were made of lead. In many ways, lead was an ideal material for this application: It is highly malleable, it has a low melting temperature, and it was inexpensive and readily available.

- Roman pipes were fabricated in standard diameters ranging from about 1 inch to nearly 2 feet. To make a pipe, the ***plumbarius*** poured molten lead into a flat mold of stone, clay, or sand. This mold was 10 feet long—the standard length of a Roman pipe—and slightly wider than the required circumference. The pipe thickness was standardized by specifying the weight of lead poured into the mold for a given pipe size.

- While this lead sheet was still hot and flexible, the *plumbarius* bent it around a cylindrical form of wood or bronze, then withdrew the form and closed the seam by folding the edges or by using **solder**. A common technique for soldering a pipe was to place temporary ridges of clay along either side of the seam to hold the molten solder in place as it cooled and solidified.

The Myth of Lead Poisoning

- No discussion of lead pipes in Rome would be complete without considering the issue of lead poisoning. Much has been written about this subject, including a rather astonishing theory that the decline of the Roman Empire was caused, not by barbarian incursions, imperial overstretch, or civil wars, but by lead contamination of Roman water systems. This claim does not stand up to scientific scrutiny.

- First, water in a Roman distribution system was constantly moving. The elapsed time between water's first entry into a lead pipe and its arrival at a public fountain could not have been more than a few minutes—not enough exposure for an appreciable amount of lead to enter the water. Furthermore, because most Roman water sources had high mineral content, a coating of calcium carbonate would have quickly accumulated on the insides of most pipes. This coating would have prevented direct contact between the water and the inside pipe surface; thus, there would have been very little opportunity for contamination.

- What's more, not all metal components of Roman water supply systems were made of lead; some connectors were made of bronze. A certain type of connector, called a ***calix***, was used to regulate water use. When a city dweller wanted to obtain his own private water supply, he arranged for the Roman water authorities to install an officially approved *calix* in the local secondary *castellum*. The customer then hired a *plumbarius* to run a pipe from the *calix* to his home—and paid his water bill based on the size of the *calix*.

Public Fountains

- Water passed through the *calix* into a supply pipe and down to a public fountain—one of which was usually located at the base of the secondary *castellum*. Public fountains served as the main source of drinking water for most city dwellers.
    - The typical public fountain was a rectangular stone basin with a decorated spout on one end and an overflow trough cut into the upper edge of one side.
    - In a Roman continuous-flow system, water was constantly overflowing from these fountains and cascading across the streets en route to the sewers—a primitive form of street cleaning.

- The other pipes exiting from the secondary *castellum* delivered water to private residences. A private water supply was evidently a major status symbol; domestic supply pipes were often mounted on interior walls and occasionally were made of silver. The supply pipe typically emptied into a basin at a central location within the house—not a kitchen or bathroom, as we might expect—and the overflow was then channeled into the sewers, often by way of a private garden.

- Although continuous flow was the norm in Roman water systems, quite a few supply lines at Pompeii incorporated faucets or taps. Some of these devices were apparently used just as modern faucets are: to turn the water on or off at a basin or fountain. But more often, they were used as shutoff valves at pipe junctions, to isolate a branch of the system for maintenance or repair.

The Sewer System

- The Roman public toilet was typically a large, open room with marble benches on three sides. The facility was served by two continuously running water channels—one beneath the seats (to carry away waste) and one just in front of them (for hand washing).

- Although not all Roman cities had well-developed sewer networks, in cities that did, these systems were often quite sophisticated. At Pompeii, masonry drainage channels with gabled covers ran underneath the sidewalks. Storm runoff and overflow from the fountains entered these channels through openings in the pavement. Other wastewater entered these channels underground, from drainpipes originating in baths, public toilets, and private homes.

- However, a great deal of wastewater (including human waste) was still dumped directly into the streets—even in cities with public toilets and well-developed sewer systems. This wastewater combined with overflow from the public fountains to produce a continuous stream of filth running across the streets and into the drain inlets.

- In well-developed Roman sewer systems, the wastewater channels emptied into larger collector channels, and these, in turn, emptied into a single "great sewer" running beneath the center of the urban

© Alago/Wikimedia Commons/Public Domain.

Tall stepping stones allowed pedestrians to keep their feet out of the wastewater than ran continuously through urban streets.

area. The great sewer collected all the city's wastewater and storm runoff and channeled it into the nearest river, where it rejoined the natural pathway of the hydrologic cycle—while also contributing to the systematic pollution of ancient waterways.

- In the next lecture, we'll visit Rome's most voracious user of water: the imperial bath—an infrastructure system that incorporated nearly every one of the Roman technologies we have discussed thus far: planning, surveying, monumental stone masonry, concrete vaulting, domes, roads, water distribution, drainage, and a few others we have not yet encountered.

Important Terms

calix: A standardized bronze fitting used to regulate the amount of water supplied to individual users in a Roman water distribution system.

castellum divisorium: A structure that divides the incoming flow from an aqueduct into multiple channels for subsequent distribution.

plumbarius: A Roman plumber.

secondary *castellum*: A structure that distributes an incoming flow of water to multiple users while also controlling pressure in the water distribution system.

settling tank: A masonry tank used to remove sand, silt, and other suspended solids from aqueduct-supplied water before it entered the urban distribution network.

solder: A mixture of lead and tin used to connect lead pipes.

Suggested Reading

Adam, *Roman Building*, chapter 10.

Hodge, *Roman Aqueducts and Water Supply*, chapters 11–12.

Landels, *Engineering in the Ancient World*, chapter 2.

Oleson, *The Oxford Handbook of Engineering and Technology*, chapter 11.

Questions to Consider

1. Why have so few elements of ancient water distribution systems survived to the present day?

2. How do secondary *castella* control pressure in an urban water distribution network?

3. Why did Roman cities require such huge quantities of water in comparison with earlier Greek and Hellenistic cities (and even modern ones)?

Paradigm and Paragon—Imperial Roman Baths

Lecture 15

The ritual of sauna, hot bath, cold plunge, and massage was an essential part of Roman culture throughout the empire. By 33 B.C., there were 170 public baths in the city of Rome alone. At the beginning of the imperial era, Roman emperors quickly recognized the value of public bath construction as a means of demonstrating their power and winning public favor. Marcus Agrippa, Augustus's right-hand man, built a major bath complex in the heart of Rome in 19 B.C. Then, Nero, Trajan, Caracalla, and Diocletian responded with successively larger and more magnificent facilities. In this lecture, we'll examine the best-preserved and best-documented of these grand public works: the Baths of Caracalla.

Precursors to the Roman Public Bath

- During the Classical era, the Greeks began constructing public bath facilities for common citizens. These typically consisted of a single room with a row of individual stone tubs along one wall. Cold water was drawn from a local spring or well and then poured over the individual bather by an attendant. By the Hellenistic era, individual baths began appearing in private homes, and public facilities became somewhat more sophisticated.

- Two fundamental characteristics of the Roman bath did not originate in the Greek world, however. These were: (1) a sequence of three adjacent temperature-controlled rooms—one cold, one warm, and one hot, and (2) communal hot-water bathing pools. The origin of these features is uncertain, but they appear to have emerged gradually in the 3rd and 2nd centuries B.C. in southern Italy, perhaps in imitation of the natural hot mineral springs.

Caracalla: A Vast Bathing Facility

- Emperor Caracalla inaugurated the Baths of Caracalla in A.D. 216, and they were fully completed in A.D. 235. The bath building was immense—larger than four football fields placed side by side. Its layout was symmetrical, with four main bathing spaces, defined below, aligned on a central axis:

- o ***Natatio***, a large open-air swimming pool.

- o ***Frigidarium*** (cold-water bath), a vast groin-vaulted hall with four cold-water plunge pools set into its corners.

- o ***Caldarium*** (hot-water bath), a circular domed structure similar to the Pantheon, with seven hot-water pools placed in arched openings around its perimeter.

- o ***Tepidarium*** (with two warm-water pools), provided a transition between the *frigidarium* and *caldarium*.

- On opposite sides of the central axis were two ***apodyteria***, or changing rooms, just inside the main entrances; two colonnaded

The Baths of Caracalla was a vast facility, capable of handling 6,000 bathers per day.

open-air exercise areas, called ***palaestrae***; and matched pairs of saunas and other heated rooms arrayed along the southwestern wall. The upper level of the building featured sundecks and promenades.

Site Preparation

- To better understand the Baths of Caracalla as an integrated infrastructure system, we will demonstrate how it was constructed. The basis for this reconstruction is a study by Janet DeLaine, a classics scholar at the University of Oxford. She divided construction of the baths into a series of specific tasks; determined their most likely sequence; and then estimated the people, materials, equipment, and time required to perform each task.

- In the spring of A.D. 212, site preparation would have begun with an army of manual laborers cutting three terraces into the slope of the Aventine Hill. A common practice in Roman construction, terracing was intended to facilitate a highly accurate site layout by starting from an absolutely level surface.

- Concurrent with site preparation, construction began on a new aqueduct, the Aqua Nova Antoniniana—a branch connecting the Aqua Marcia to a huge cistern on the southwestern side of the bath complex. The aqueduct and cistern were high-priority construction tasks because mixing concrete for the remainder of the project would require huge quantities of water.

- The cistern was a rare exception to the normal mode of continuous flow in Roman water systems. Because the Baths of Caracalla would be such a voracious consumer of water, substantial storage capacity was required to avoid disrupting the water supply to the city.

Structural Foundation

- To construct the structural foundation, wherever there was a wall in the design, a trench was dug—22 feet deep for the *frigidarium* and *caldarium* and 15 feet deep for the remainder of the complex. Workers then installed the main sewer, which was a vaulted brick gallery, positioned just below the surface of the terrace.

- The structure was built on a platform, 26 feet above the terrace, to make room for an intricate network of subterranean infrastructure: water supply passages, drainage structures, and maintenance galleries. This platform was composed of four main elements:
  - Substructures made of thick walls of *opus testaceum*, extending upward 26 feet from the foundations of the bath building.
  - A two-level arcade and a pair of curved retaining walls that enclosed the front and sides of the platform.
  - An multilevel network of passageways interwoven with the substructures.
  - A huge mass of fill—a mixture of compacted soil, pozzolana, and brick fragments—that surrounded and supported the underground infrastructure and created a level working surface for the bath building's main floor.

Construction of the Bath Building

- With the platform complete, construction of the bath building began. First, the floors of the *frigidarium*, *natatio*, and *palaestrae* were reinforced with 2-foot-thick concrete slabs. Subfloor elements, such as stone column bases and the 5-foot-deep *frigidarium* pools, were also installed at this time.

- Next, the walls were erected, to a height of nearly 80 feet for most of the building. This work required about 4,000 masons and laborers—and took most of the year A.D. 214 to complete. Like the substructures, these walls were *opus testaceum*—though here, the concrete aggregate was lightweight tufa, rather than basalt.

- There was extensive use of relieving arches, which were embedded in the walls to divert the weight of the structure around doorways, windows, and niches. The walls also included built-in roof drainpipes, channels for water supply pipes and heating flues, internal staircases, and mounting sockets for decorative stone elements.

- Meanwhile, carpenters prefabricated wooden centering for an incredible variety of concrete barrel vaults, groin vaults, semi-domes, and domes. Most of these vaults were built using a technique that had become quite common in imperial-era construction. The centering was covered, not with wooden planking, but with a layer of heavy terra-cotta bipedales. When concrete was poured on top, it adhered to the bricks but not to the wood—greatly facilitating removal of the centering after the concrete had cured.

Design of the Interior

- With the vaulting in place, interior construction could proceed under cover, protected from the elements. A total of 252 granite columns were installed throughout the building—to delineate the *palaestra* porticos, to frame doorways, and to screen passages between adjacent rooms. These columns were used solely for architectural effect; the real structural load bearing was done by the brick-and-concrete arches.

- Even the spectacular 50-foot granite columns adorning the *frigidarium* were nonstructural. The *frigidarium* vaulting was actually supported by stone impost blocks, which were embedded deeply into the massive brick-and-concrete piers behind the columns. The immense vertical and horizontal forces generated by the vaults were transmitted through the impost blocks and into the piers.

- The bath building's most impressive structural elements were the great concrete roofs of the *frigidarium* and *caldarium*. The *frigidarium* roof consisted of three groin-vaulted bays spanning nearly 80 feet across the central hall. The structure of the *frigidarium* was beautifully integrated with its function. The coffered vaulting beautifully enclosed this majestic space, while providing large upper-level windows for natural illumination.

- The *caldarium* dome was an adaptation of the Pantheon's. Although the *caldarium* dome was somewhat smaller in diameter, it was equal in height and more structurally daring, because the drum was pierced by large windows on two levels. These would have lightened and weakened the walls, significantly increasing the challenge of resisting the dome's outward thrust. As they had done in the Pantheon, engineers of the *caldarium* met this structural challenge with substantial radial buttress vaults embedded within its walls.

An Effective Heating System

- Among the last functional elements added to the bath building were the **hypocausts**.
  - A hypocaust was a heating system composed of three principal components: a wood-burning furnace, or ***praefurnium***; a raised floor, created by placing several layers of large bricks on stacks of smaller bricks; and a series of rectangular terra-cotta pipes, called ***tubuli***, embedded in the walls of the room.
  - In operation, the *praefurnium* heated the air beneath the raised floor and because hot air rises, it was drawn up through the *tubuli* and vented out of the building. As the hot air circulated, it heated both the floor and the walls quite effectively.
- The building's layout reflects a surprisingly sophisticated appreciation for energy efficiency. The *caldarium* and other heated rooms were situated in a single row facing southwest because these rooms would receive direct sunlight from midday until evening, when the baths were most heavily used.
- Furthermore, the designers of the Baths of Caracalla were able to maximize the effects of solar heating by taking advantage of significant imperial-era improvements in glass-making technology. The windows of the rooms were glazed to reduce heat loss, while also augmenting hypocaust heating through the greenhouse effect.

Integrated Infrastructure at Its Finest

- The imperial bath complex is integrated infrastructure at its best—one of the ancient world's finest examples of an engineered system. Of many noteworthy aspects of the baths' design, none is more impressive than its water supply system.

   - A cistern holding 3 million gallons was filled during off-peak hours by the Aqua Nova Antoniniana.

   - Then, during operating hours, water flowed continuously from this reservoir through a network of lead pipes to various pools, fountains, and cauldrons throughout the building. Many of these pools communicated with others through waterfalls.

- Bronze cauldrons were used to heat water for the *caldarium* pools, and after it had cooled a bit, this water was channeled into the *tepidarium* pools before being discarded. Some of the wastewater was used to continuously flush the public toilets, and some was even used to drive a waterwheel-powered grain mill located in an underground chamber. All the wastewater eventually found its way into an extensive system of underground drainage structures and then into the Tiber.

Important Terms

apodyterium: A room for changing clothes in a Roman bath.

caldarium: The hot room in a Roman bath.

frigidarium: The cold room in a Roman bath.

hypocaust: The system used to heat both water and air in a Roman bath.

natatio: A Roman open-air swimming pool.

palaestra: A Roman open-air exercise area.

praefurnium: The wood-burning furnace that supplied heat in a hypocaust.

tepidarium: The medium-temperature room in a Roman bath.

tubuli: Terra-cotta pipes used to heat the walls in a hypocaust system.

Suggested Reading

DeLaine, *The Baths of Caracalla.*

Malacrino, *Constructing the Ancient World*, chapter 9.

Questions to Consider

1. How does the layout of the Baths of Caracalla reflect Roman urban planning principles?

2. What characteristics of the Baths of Caracalla reflect a comprehensive, well-integrated design that was completed before the start of construction?

3. To what extent does this structure reflect the products of the Roman construction revolution?

Harnessing Animal Power—Land Transportation

Lecture 16

In classical antiquity, there were four principal sources of power: humans, animals, water, and wind. Human power was used primarily for specialized tasks that required brainpower and dexterity. Animals were used primarily for pushing or pulling. Water power was used to a lesser extent than human or animal power, primarily for milling grain and pumping water, and wind power was used almost exclusively for sailing ships. In today's world, it is difficult to appreciate how meager these ancient power resources were. Yet as we'll see in the next eight lectures, ancient engineers often responded to this constraint with extraordinary ingenuity—developing machines that achieved impressive levels of performance using power inputs that, by modern standards, were quite miniscule.

Six Simple Machines

- A machine is an assembly of fixed or moving parts used to perform work. In this context, "work" has a precise scientific definition: **Work** is the quantity of energy expended when a force moves through a distance. We often evaluate machines in terms of **power**—defined as the rate at which work is done.

- To understand complex machines, it is helpful to analyze the system as an assembly of simple machines, each performing a specific function. We owe this concept to the great 3rd-century-B.C. mathematician Archimedes, who identified three simple machines:
  - The **lever**, which magnifies an applied force.

  - The **pulley**, which changes the direction of a force.

  - The **screw**, which converts a rotational force (or torque) to a linear force.

- Later Greek thinkers added three more simple machines:

- The **wedge**, which converts a single force into a pair of opposing forces that are oriented perpendicular to the surfaces of the wedge.

- The **inclined plane**, which allows an object to be lifted with an applied force less than the object's weight.

- The **wheel and axle**, which facilitates horizontal movement with a minimum application of force.

- Although the six simple machines were first identified as such during the Classical era, all but the screw were well known before that time. One of these simple machines—the wheel and axle—is particularly important to our understanding of land transport technology.

Types of Land Transport

- In the ancient Mediterranean world, long-distance transportation of heavy cargo was done almost exclusively by ship. Land transport was limited by its inherent expense and, before the Romans came along, by poor-quality roads. The great Roman network of roads was built primarily to facilitate the movement of infantry, however, not vehicular traffic. Throughout antiquity, there were three principal modes of land transport: human porters, pack animals, and wheeled vehicles pulled by draft animals.

- Human porters could carry loads of 50 to 60 pounds for relatively short distances. For example, on a construction project, such as the Baths of Caracalla, the required construction materials would have been delivered to the site by ox-drawn wagons, but the bricks, stone, pozzolana, and timber would all have been moved within the site by human muscle power.

- Pack animals carried heavier loads over longer distances. A donkey might carry as much as 200 pounds, depending on its size, and a mule, 300 pounds. In the Greek and Roman worlds, pack animals were often provided by transport contractors, who maintained large fleets of animals and hired them out for specific jobs.

- To move people and heavy loads for long distances overland, the preferred mode of transport was a wheeled vehicle pulled by a draft animal.
    - Some common types of two-wheeled vehicles included the Greek war chariot; the Greek ***hamaxa***, a small all-purpose carriage; the famous Roman racing chariot; and a Roman two-seat taxi cab called the ***cisium***. All were light, fast, and maneuverable but had relatively low load-carrying capacity.

    - Four-wheeled vehicles were less maneuverable but could carry significantly more cargo. Their configurations ranged widely and included crude, solid-wheeled ox carts, used extensively for hauling timber and stone; specialized cargo carriers, such as a wine wagon; and extraordinarily sophisticated Roman traveling coaches.

The Wheel: A Hallmark of Human Civilization

- All ancient vehicles incorporated three principal technologies: wheel, axle, and traction system. One of the great hallmarks of human civilization, the wheel was probably invented around the 4th millennium B.C. By the 2nd millennium B.C., the state of the art had advanced considerably, as evidenced by the use of sophisticated, lightweight spoked wheels on military chariots in the Near East. Wheels of a variety of forms were still in widespread use during the Classical era.

- Solid wooden wheels were used on vehicles ranging from ox carts to massive siege engines, where simplicity and strength were the governing design criteria. The Greek *hamaxa* used unique crossbar wheels that were very light.

- During the Roman Empire, the spoked wheel reached a high level of technological sophistication.

- The fully developed Roman wheel had as many as 12 elegantly crafted spokes, fitted into a wooden hub that was reinforced with iron bands to prevent the wood from splitting under heavy loads.

- Its rim, or ***felloe***, was fabricated from a single piece of ash wood, bent into a circle, spliced together, and then enclosed within an iron tire—which strengthened the wheel while also preventing wear.

- The iron tires found in archeological digs have no nail holes, which tells us that they were shrink-fitted into position. After the fall of Rome, this technology would not be equaled again until the High Middle Ages.

The Axle

- A wheel is of little use without an axle, which is probably why the Greeks defined the wheel and axle, together, as a simple machine. There were two basic axle configurations. In a fixed axle, the wheels rotated on the axle. In a second configuration, the wheels were fixed to the axle, and the axle rotated.

- The rotating-axle configuration tended to be used only for simple farm carts, while the fixed axle was used for chariots and other vehicles requiring high speed and maneuverability. To address the problem of excess wear on fixed axles, the Romans developed an iron wheel bearing system—with one cylindrical sleeve fitted over the axle and a slightly larger one reinforcing the inside of the wheel hub.

- A two-axle wagon used either fixed or rotating axles. Even the more advantageous fixed-axle configuration (with all four wheels rotating independently) does not prevent a wagon from skidding when pulled around a curve, however. The only way to address this problem is with a pivoting front axle, where the front axle rotates on heavy iron pin; a draft pole is rigidly attached to the axle.

- Evidence of wagons with a pivoting front axle have been found in Roman-era Thrace. A Roman traveling coach, or ***carruca***, was stunning evidence of the technological sophistication achieved by imperial wagon builders. Not only did the coach have a well-designed pivoting axle and fine spoked wheels with iron tires and bearings, but it also featured a full suspension system.

The suspension system of this imperial-era *carruca* would have greatly enhanced the comfort of the coach's occupants as they rolled across the uneven paving stones of the Roman roads.

The Traction System

- The third principal technology found in ancient vehicles was the traction system, consisting of the draft animals and the apparatus that connected them to the vehicle.

- With its high power output and docile temperament, the ox was the ancient world's engine of choice for pulling heavy wagons. Oxen were typically used in pairs, and they applied tractive force through a **yoke**, which was connected to the vehicle by a single draft pole. The yoke was developed in the Neolithic Age and is still used in many parts of the world today. It consists of a heavy wooden beam fitted over the animals' shoulders and held in place with a harness.

- The ox's principal limitation was its low speed—roughly 1 mile per hour. For such vehicles as the *hamaxa* and the *cisium*, which carried light loads and had a "need for speed," donkeys and mules were generally used for pulling. These animals' skeletal structures were also fairly well suited for the yoke; thus, the traction systems used would have been similar to the ox yoke.

The Role of the Horse in Antiquity

- Horses were almost never used as draft animals in the Classical world. There is considerable scholarly disagreement about why this was the case. One popular theory holds that the horse was ineffective as a draft animal because its skeletal structure was unsuitable for the yoke. Because a horse holds its head upright and lacks pronounced shoulder blades, the yoke had to be cinched tightly around its neck. When the animal attempted to pull a heavy load, the harness restricted its breathing, greatly reducing its maximum pulling force.

- This theory has been largely disproved by modern scholars. Archeological discoveries and experimental reconstructions have demonstrated conclusively that the Romans developed an effective horse yoke that used the animal's chest for traction—thus permitting the horse to pull without strangling itself.

- If a suitable yoke existed, the question remains: Why weren't horses used more often as draft animals? The likely answer is partly cultural and partly economic.
    - From a cultural perspective, the horse occupied an exalted position in the Classical imagination. It was seen as the mount for gods and heroes—not a beast of burden.

    - From an economic perspective, horses were significantly more expensive to own than donkeys, mules, or oxen. As evidence, consider that the second highest of the four Athenian social classes was called the *hippeus*, which referred originally to privileged men who were wealthy enough to own horses and fight in the cavalry. The horse was not just expensive to buy, but it was also expensive to feed.

- Indeed, the horse's only distinct advantage as a means of land transport was its superior speed. And that is why, in antiquity, horses were used almost exclusively in the few situations where speed mattered and cost did not: as cavalry mounts, in chariot racing, and in state-sponsored courier services, such as the Persian Royal Road system and the Roman *cursus publicus*.

Important Terms

carruca: A large four-wheeled Roman traveling coach.

cisium: A Roman two-wheeled carriage, used as a taxi.

felloe: The outer rim of a spoked wheel.

hamaxa: A small two-wheeled, all-purpose carriage used by the Greeks.

inclined plane: A simple machine that allows an object to be lifted with an applied force less than the object's weight.

lever: A simple machine that magnifies an applied force.

power: The rate at which work is done.

pulley: A simple machine that changes the direction of a force.

screw: A simple machine that converts a rotational force (or torque) to a linear force.

wedge: A simple machine that converts a single force into a pair of opposing forces that are oriented perpendicular to the surfaces of the wedge.

wheel and axle: A simple machine that facilitates horizontal movement with a minimum application of force.

work: The quantity of energy expended when a force moves through a distance.

yoke: A heavy wooden beam fitted over the shoulders of a draft animal and held in place with a harness.

Suggested Reading

Landels, *Engineering in the Ancient World*, chapter 7.

Oleson, *The Oxford Handbook of Engineering and Technology*, chapter 23.

Weller, "Roman Traction Systems."

White, *Greek and Roman Technology*, chapter 10.

Questions to Consider

1. Why were humans often used for power in the ancient world, even though animals could produce significantly higher power output?

2. What devices do you currently own that are capable of producing more sustained power than an average human (approximately 1/10 horsepower)?

3. Why did significant advances in Roman land transport technology *not* result in the elimination of older, less sophisticated transport systems?

Leveraging Human Power—Construction Cranes

Lecture 17

In this lecture, we'll look at the ingenious wood-and-iron construction cranes that were used to build many of the ancient world's greatest structures. We'll explore how these fascinating human-powered machines worked by significantly amplifying the force exerted by human muscles. In a broader sense, we'll understand how they exemplified the ingenuity of the ancient engineer. Why were such machines needed when Greek and Roman builders could use vast numbers of slaves to do the heavy lifting on a construction site? The answer is that there were many types of work that simply could not be performed by unassisted human power, no matter how much of it was available.

The Crane: More Practical and Economical than Human Power

- To understand the importance of construction machines, consider the fact that lifting a 10,000-pound stone block would take 100 slaves. However, only 15 men fit around the circumference of the block. Clearly, it doesn't matter how many slaves are available—the job simply cannot be done by human power without some form of mechanical augmentation.

- To lift a column, an engineer might consider an inclined plane, which works by trading force for distance. The lower the pushing force, the longer the distance through which that force must be applied. However, if the stone columns adorning the upper level of the Colosseum had been raised to their required 160-foot height with a 10-degree ramp, that ramp would have to have been nearly 1,000 feet long—significantly longer than the Colosseum itself.

- The construction crane provided a more practical and economical means of lifting heavy objects. Although no such machines have survived from antiquity, we have ample evidence of their existence from as early as the 6th century B.C.—in sculpted reliefs, in unfinished stone blocks, and most important, in Book X of *De*

Architectura, where Vitruvius provides excellent descriptions of several standard crane configurations.

Shear-Leg Crane

- The shear-leg crane consisted of two wooden posts, connected with iron brackets and supported by backstays. A lifting rope was pulled by a **windlass**, a device that combined two simple machines: the shaft (wheel and axle) and handspikes (levers). It was operated by two men, alternately inserting handspikes into hubs and pulling.

- With each pull of a handspike, the crane operators applied **torque** to the shaft. Torque is defined as the tendency of a force to cause rotation. To calculate torque, multiply the force by its distance from the center of rotation; thus, to increase torque (and, therefore, to increase the shaft's tendency to rotate), one can either increase the force or increase the distance.

- The windlass was used in this type of crane because it provided **mechanical advantage**—the amplification of an applied force by a mechanical device. For every 1 pound applied to the handspike, 20 pounds of force were developed in the lifting rope. The windlass achieves mechanical advantage by trading force for distance—just as the inclined plane does.

- It is important to note that some of the work input is actually lost to friction in the axle bearing. Thus, the machine designer needed to minimize friction in order to maximize work output. That's why the Romans used precisely fabricated iron wheel bearings on their wagons.

Block-and-Tackle Pulley System

- The crane also incorporated another simple machine: the pulley. More precisely, it used a **block-and-tackle system**, which provides mechanical advantage through the use of two pulleys rather than one. The upper pulley was a double block (two sheaves mounted side by side); the lower pulley was a single block.

- By combining multiple pulleys in this way, the block-and-tackle system achieves a mechanical advantage of three to one; in other words, a 1-pound pull on the rope will lift 3 pounds. Once again, mechanical advantage is achieved by trading force for distance. To lift the load 1 inch, a worker needs to pull 3 inches of rope through the pulley system.

- The crane was a superb example of a **compound machine**: a system composed of multiple simple machines. The mechanical advantage of a compound machine is equal to the product of the individual mechanical advantages of its components. Thus, for this crane, the total mechanical advantage is 20:1 (for the windlass) × 3:1 (for the block-and-tackle) = 60:1. Even with the inefficiencies of friction taken into account, this machine would still allow a single operator pulling with 100 pounds to lift more than 2 tons.

A More Powerful Roman Crane

- The cranes described by Vitruvius were originally developed by the Greeks and subsequently adopted by Roman builders. The construction crane that emerged during the Roman Empire was essentially identical to the Greek shear-leg crane, except that it was much larger and more powerful, with greater lifting capability.

- Its mechanism for harnessing human power was the **tread wheel**—a large wooden spoked wheel attached to the crane's driveshaft. It was turned by one or more men walking inside the wheel. The Roman tread wheel provided far more torque than the Greek windlass, because its radius was significantly larger and its operators were able to apply their full body weight to the wheel.

- The crane's large lifting block had three sheaves—which typically would be paired with a two-sheave block below to produce a five-to-one mechanical advantage for the lifting apparatus alone. The mechanical advantage of the system would have been greatly increased by the huge tread wheel, which was large enough to accommodate five crewmen. Based on all these characteristics, scholars have estimated that this crane could lift about 8 tons to a height of perhaps 40 feet.

Machines Tailored to Unique Projects

- As capable as these machines were, there were certain demanding and unique construction challenges for which standard cranes were entirely inadequate. In such cases, special high-capacity lifting devices would have been specifically designed to meet the express demands of the project.

© iStock/Thinkstock.

Trajan's Column is famous for its intricately carved spiral frieze illustrating the emperor's military victories, but it should be equally famous as one of the most extraordinary construction achievements in antiquity.

- Trajan's Column in Rome is a fine example. It was 126 feet tall, measured from ground level to the base of the bronze statue of Trajan that once stood at its apex. Above its 20-foot-square pedestal, the shaft is composed of 20 solid marble drums. The heaviest of these drums—the base—weighed 60 tons. The uppermost drum, which included the column's capital and observation platform, was only slightly lighter—59 tons—and it had to be lifted more than 110 feet to the top.

- The standard Roman construction crane—robust as it was, with its 8-ton capacity and 40-foot height—was entirely inadequate for this unique project. No one knows how Trajan's Column was built, but Lynne Lancaster, a classics scholar at Ohio University, has provided a well-researched and well-reasoned construction scheme.
  - This scheme posits a massive wooden tower surrounding the column pedestal and providing structural support for lifting the massive marble drums.

  - It certainly makes sense that Apollodorus of Damascus would have used a temporary wooden tower to build Trajan's Column. He was, after all, an accomplished military engineer who would have been familiar with wooden siege towers of comparable size. Indeed, he even wrote a treatise on siegecraft that included technical descriptions of such towers.

  - Lancaster's proposed construction tower consists of two parallel shafts—one for lifting the column drums and one for lowering them into position on the column. The drums are moved on rollers. Given the 60-ton weight of the heaviest drum and the 8-ton lifting capacity of a Lewis bolt, eight sets of lifting ropes would be required to raise the drums. If the lifting ropes incorporated block-and-tackle systems, the rope tension would be reduced significantly below 8 tons.

  - The block-and-tackle systems are suspended from heavy timber trusses spanning 22 feet across the top of the tower. Trusses are essential here, because wooden beams of this length could not possibly have carried such heavy loads.

  - The lifting ropes are routed from the upper blocks down to pulleys at the base of the tower and then out to a battery of eight **capstans**. A capstan is basically a horizontally oriented windlass anchored to the ground. But unlike a windlass or tread wheel, a capstan can be turned by either men or animals—and many more "pushers" can be brought to bear simultaneously, resulting in significantly greater torque.

- Also, the capstan allows the application of torque to be more precisely controlled—a vital feature for the reconstruction of Trajan's Column because all eight lifting ropes must carry equal shares of each drum's weight. If one rope carried significantly more than its fair share, it would probably snap or its Lewis bolt would fail—and then a chain reaction failure of the other lifting ropes would ensue.

- Once a column drum is hoisted to its required height, it must be moved horizontally from the lifting shaft to the lowering shaft—another daunting challenge. In Lancaster's reconstruction, this movement is accomplished by using a temporary floor that is moved out of the way as the drum is lifted above it, then lowered into position to support the horizontal movement of the drum on rollers, and then moved out of the way again as the drum is lowered into its final position on the column. The drum is locked into position with iron dowels (in the traditional Greek manner).

- Whether or not this reconstruction is accurate, what we do know with certainty is that Trajan's Column was built successfully. That achievement required the design and construction of a mechanical system capable of repeatedly lifting 60-ton loads, moving them horizontally, and lowering and positioning them with great precision, high above the ground, on a tightly confined site, using ropes and fixtures of limited capacity, propelled solely by human or animal power. However it was done, the building of Trajan's Column was a tour de force of technological skill.

Important Terms

block-and-tackle system: A system of pulleys and ropes, which provides mechanical advantage for lifting.

capstan: A device for harnessing and amplifying human or animal power, to apply tension to a rope. A capstan has a vertical shaft, while a windlass typically has a horizontal shaft.

compound machine: A system composed of multiple simple machines.

mechanical advantage: The amplification of an applied force by a mechanical device.

torque: The tendency of a force to cause rotation, expressed in terms of force times distance.

tread wheel: A device for harnessing and amplifying human power in construction cranes, water-lifting devices, and similar machines.

windlass: A device for harnessing and amplifying human power, to apply tension to a rope. A windlass typically has a horizontal shaft, while a capstan has a vertical shaft.

Suggested Reading

Lancaster, "Building Trajan's Column."

Landels, *Engineering in the Ancient World*, chapter 4.

Oleson, *The Oxford Handbook of Engineering and Technology,* chapter 13.

Questions to Consider

1. What is the relationship between work and mechanical advantage? (Hint: Mechanical advantage always involves trading force for distance.)

2. Why was the Roman tread wheel–powered crane able to generate greater lifting force than earlier Greek windlass-powered cranes?

3. Do you agree with Lynne Lancaster's proposed system for the construction of Trajan's Column? What other schemes might have worked?

Lifting Water with Human Power

Lecture 18

Because of its critical role in irrigation, mining, water supply, maritime commerce, and firefighting, the water-lifting machine was an important contributor to economic development in the ancient world. In this lecture, we'll examine a number of these devices, including the screw pump, which was the first known use of the simple machine called the screw, the tympanum, the bucket wheel, the bucket chain, the *noria*, the animal-powered *saqiya*, and the force pump.

The Screw Pump

- Traditionally attributed to Archimedes in the 3rd century B.C., the **screw pump** was called a *cochlias* in Greek because it resembled the spiral seashell of the same name. Vitruvius gives us a rich description of the construction of a screw pump.

- The machine consisted of a rounded wooden rod with a length 16 times its diameter. Radial lines divided the rod's circular end into eight equal segments. A series of circles was drawn along the length of the rod at the same spacing, creating a grid of perfect squares over the entire surface of the rod.

- Engineers coated a narrow strip of flexible wood with pitch for waterproofing and wrapped it around the shaft. The resulting spiral shape is called a **helix**; because it runs diagonally across the square grid, it is said to have a helix angle of 45 degrees.

- Other strips of wood were added on top of the first until the outer diameter of this helical vane was twice the diameter of the wooden shaft. An additional vane was created for each of the remaining grid points, and then all eight vanes were enclosed in a case made of wooden slats and reinforced with iron bands.

- The screw pump lifted a continuous stream of water as long as the shaft was rotating. It has been estimated that a typical 8-foot-long screw pump was capable of pumping about 2,000 gallons per hour when powered by one man. This number—2,000 gallons per hour—is called the **flow rate**, a useful measure of a pump's output.

The Tympanum

- Because the maximum practical lift for a typical 8-foot-long screw pump would only be about 4 feet, there was still a powerful stimulus for the development of a more effective water-lifting device: the waterwheel.

- The type of water-lifting wheel called the **tympanum** had a hollow wooden cylinder that rotated on a horizontal axle and was subdivided into eight wedge-shaped compartments. The outer rim had one inlet slot per compartment; outlet holes were cut into the apex of each compartment. A wooden trough (launder) was mounted alongside the drum to catch water.

- A 10-foot-diameter, 8-inch-wide tympanum would be capable of lifting more than 5,000 gallons per hour with one man treading the rim. But, like the screw pump, the tympanum also had the disadvantage of relatively low lift.

The Bucket Wheel

- The solution to low lift was the **bucket wheel**. This device worked on the same principle as the tympanum, except that the water was carried in smaller compartments, placed around the circumference of a larger wheel. Its principal advantage was that the outlets were closer to the rim of the wheel; thus, the launder could be placed much higher—significantly increasing its lift.

- A modern reconstruction of the bucket wheel has a flow rate of about 1,200 gallons per hour—considerably less than both the screw pump and the tympanum; however, it was able to lift water about 12 feet—a threefold improvement over both. The power required to operate it was slightly more than 0.1 horsepower—indicating that the machine was perfectly proportioned to be operated by one man treading its outer rim continuously.

- The Romans also combined multiple bucket wheels to lift large quantities of water to amazing heights. Water at the Rio Tinto mines was drained by an incredible installation of eight pairs of wheels arranged in ascending steps carved into solid rock. The system had a total lift of nearly 100 feet and could move about 2,400 gallons per hour, with 16 men providing the power.

The Bucket Chain

- An alternative technology for achieving these sorts of high lifts was the **bucket chain**. According to Vitruvius, this device consisted of a series of small bronze buckets attached to two loops of iron chain. The chains were driven by a tread wheel driving a rotating axle on top; as the axle rotated, the buckets were continuously filled, lifted, and then dumped into a launder.

- The remains of a Roman bucket chain dating from around A.D. 100 were discovered in the heart of London. From the artifacts, a team of scholars, engineers, and craftspeople organized by the Museum of London were able to reconstruct a full-scale working replica of the machine.

- The chain of buckets was driven by an eight-sided wooden wheel, with projecting radial partitions engaging bent-iron links between buckets—and effectively preventing the chain from slipping. Each of these partitions also formed one side of a lateral trough, which captured water dumped from the buckets and channeled it out sideways to a launder. This ingenious system maximized the capacity of the buckets and minimized spillage.

- The Museum of London's engineers calculated the flow rate of this marvelous machine at an impressive 1,900 gallons per hour. But that number is uncharacteristically large, because these wooden buckets were twice the size of the bronze ones described by Vitruvius, because the London machine lifted water only about 10 feet, and because the modern reconstruction is powered by four men turning a windlass, rather than one man driving a tread wheel.

The *Noria*

- The flow rate of a bucket chain is heavily influenced by its power input and height of lift. Devices that can move large quantities of water generally cannot lift very high. Devices that can lift to greater heights typically have lower flow rates. At the heart of this tradeoff is a simple principle of physics: power = force × distance ÷ time.

- In order to increase the height of lift without increasing the power input, one must either reduce the amount of water lifted or increase the time required to lift it. In either case, the flow rate (in gallons per hour) must decrease. Thus, physics tells us that the inherent price of increased lift is decreased output.

- An obvious way around this physical constraint is to increase the power input. However, on many of these devices, it would have been difficult to add a second man because there simply was not enough space to do so.

- But Greek and Roman engineers did occasionally achieve the same end by using other power sources. One option was to add vanes to the outer rim of a bucket wheel so that water power could be used to drive the wheel. The resulting device, called the ***noria***, was used as early as the 2nd century B.C.

- The *noria* required a fast-moving stream. Water pressure on the vanes rotated the wheel, even as some of the water was scooped into buckets mounted on its rim. As these buckets neared the top of the wheel, they emptied their contents into an aqueduct, which was used to supply the local irrigation system. The 60-foot lift of a *noria* wheel was an indicator of the immense potential inherent in water power.

The *Saqiya*

- Ancient engineers also used animal power for water-lifting wheels. Here, the challenge was to translate the animal's horizontal pull into rotary motion around a horizontal axle. The resulting machine was the ***saqiya***.

- The *saqiya* was a modernized version of the tympanum, made of metal instead of wood and incorporating large scoops to capture more water per revolution. It was powered by a crude capstan—a pair of cattle hitched to a pole, which turned a vertical drive shaft. This shaft drove a right-angle gearbox—an important technology for water power—that rotated the tympanum to lift water for an irrigation system.

The Force Pump

- Without question, the ancient world's most sophisticated water-lifting device was the **force pump**. The force pump was described by Vitruvius in *De Architectura*, and archeologists have found numerous examples—most frequently in shipwrecks and at the bottoms of wells.

- A typical force pump was made entirely of bronze and consisted of two cylinders, connected by right-angle pipes to a central chamber, which had a single vertical outlet pipe projecting from its upper end. Each cylinder held a closely fitted piston, linked to a rocker arm by a hinged connecting rod. The rocker arm was a hand-operated lever that moved one piston up and the other down with each stroke. The outlet pipe was connected to a pivoting nozzle, indicating that the device was clearly intended for spraying.

- The pump was placed with its inlets below the water level. When the rocker arm was rotated clockwise, the piston was pulled to the top of its cylinder. The resulting suction pulled the inlet valve open and drew water into the cylinder. This suction also pulled the outlet valve shut to prevent any water in the central chamber from being sucked back into the cylinder. When the rocker changed direction, the piston was driven downward. The pressure forced the inlet valve shut and the outlet valve open, and it drove the water out of the cylinder, into the central chamber, and upward through the outlet pipe.

- The other cylinder operated in exactly the same way but on the opposite cycle. As one cylinder drew water in, the other forced it out—Thus, there was a nearly continuous flow of water up through the outlet pipe. The flow rate was estimated at roughly 180 gallons per hour.

- Perhaps the most fascinating examples of ancient force pumps, however, were the ones at the opposite end of the technological spectrum. In the 1st century A.D., the force pump was reengineered in wood. Because of its substantially lower cost, the wooden pump was produced in large numbers and used extensively at rural villas and farms. This was Roman ingenuity and pragmatism at its best.

Important Terms

bucket chain: A human-powered water-lifting device consisting of a string of buckets fixed to a pair of iron chains, driven by a rotating shaft.

bucket wheel: A human-powered water-lifting device consisting of a rotating wheel with buckets positioned around its outer rim.

flow rate: The output of a water-lifting device, expressed in volume per unit of time (e.g., gallons per hour).

force pump: A human-powered water-lifting device that uses one or more pistons, cylinders, and valves to pump water through reciprocating (up-and-down) motion.

helix: The spiral shape associated with a screw.

noria: A water-powered bucket wheel.

saqiya: An animal-powered tympanum.

screw pump: A human-powered water-lifting device consisting of one or more helical vanes rotating on an inclined shaft.

tympanum: A human-powered water-lifting device consisting of a compartmented drum rotating on a horizontal shaft.

Suggested Reading

Landels, *Engineering in the Ancient World*, chapter 3.

Oleson, *Greek and Roman Mechanical Water-Lifting Devices*.

———, *The Oxford Handbook of Engineering and Technology*, chapter 13.

Questions to Consider

1. Why was water lifting the stimulus for so much technological development in the ancient world?

2. Explain the inherent tradeoff among power input, flow rate, and height of lift for a water-lifting machine.

3. Why do you think the Romans preferred the bucket wheel to the tympanum for draining mines?

4. Why was the force pump used most often for lifting water from wells and pumping bilgewater from ships?

Milling Grain with Water Power

Lecture 19

In this lecture, we'll look at a source of power with the greatest potential to do useful work: water power. Interestingly, in the ancient world, water power was used almost exclusively for one activity: milling grain. At that time, cereal grains—primarily wheat and barley—provided 70 to 75 percent of the calories in the average person's diet. Before grain can be consumed, however, it must be processed into flour by cracking open its outer husk, or chaff, and then grinding the inner portion into a fine powder. Throughout much of human history, grinding grain was an arduous, tedious, and time-consuming aspect of everyday life. It was a human activity that was ripe for technological improvement.

Early Grain Mills

- From the Neolithic Age to the Classical era, the standard technology for milling grain was the **saddle quern**—a simple device consisting of two shaped stones. The miller pushed the upper stone back and forth repeatedly to crush the grain. The **hopper mill** was a minor product improvement over the saddle quern; it still used two stones, but the upper stone was moved across the lower with a lever.

- A **rotary quern** was constructed of a dome-shaped lower stone, a concave upper stone with a hopper carved into its center, and a wood or iron pivot, called the **spindle**, on which the upper stone rotated. A rotary quern was more portable and easier to operate than the hopper mill.

- Sometime around the 3rd century B.C., animal-powered versions of the rotary quern began appearing in Italy. This development seems to have coincided with the emergence of commercial bakeries in many towns and the consequent need for large-scale production of flour. The fully developed form of this machine is known as the **Pompeiian mill**, because many were discovered in the ruins of Pompeii.

- The Pompeiian mill had two main components, both typically carved from solid basalt.
  - The cone-shaped lower part (the ***meta***) held a spindle on which the upper part (the ***catillus***) rotated. The lower half of the *catillus* fitted closely over the *meta*, and its upper half served as a hopper for raw grain.
  - The miller filled the hopper with grain and yoked his donkeys to the capstan arms. As the animals walked around the mill, the rotation caused grain to sift down into the gap between the *catillus* and the *meta*, where it was ground between the two stone surfaces.

Using a quern to grind grain required about three hours to produce enough flour to feed one household for one day.

- The Pompeiian mill liberated humans from the drudgery of grinding grain by hand. More important, it demonstrated that a new power source could produce dramatic improvements in both productivity and the consistency of the milled product. In that sense, it set the stage for the revolutionary development of water-powered mills.

- Recent archeology suggests that water mill technology was invented around the 3rd century B.C. and, by the 1st century A.D., had been widely adopted across the Roman world. These mills were driven by three different types of waterwheels: the **undershot wheel**, the **overshot wheel**, and the **vertical-shaft wheel**, each with its own unique advantages and disadvantages.

The Undershot Wheel

- In an undershot wheel, the wheel shaft was horizontal, and the upper millstone turned on a vertical spindle. The system included

right-angle gearing to convert horizontal shaft power into vertical shaft power.

- The principal advantages of the undershot wheel were its simplicity and low cost. The wheel could be installed along the shore of a fast-moving river or stream, and no any special structures were required to bring water to the wheel or to carry it away. But this wheel also had several significant disadvantages. It required a relatively large volume of fast-moving water to turn and, even then, its power output was relatively low.

- Moving water exerted a force on the submerged vanes of the wheel. As that force moved a given vane forward, it generated power. This force caused the wheel to rotate, but as soon as the vane started moving, the force of the water pushing on it decreased. An actual undershot wheel always moved more slowly than the water that was driving it.

- This is the underlying reason that undershot wheels have such low power output—because movement of the wheel decreases the force available to produce power. Thus, undershot wheels are inherently inefficient.

The Overshot Wheel

- The need to obtain higher power output from a smaller quantity of water was probably the stimulus for the subsequent development of the overshot wheel.

- The overshot wheel differed from the undershot wheel in two ways. Where the undershot wheel used vanes, the overshot wheel used bucket-like compartments around the rim. Where the undershot wheel was placed in the stream, the overshot wheel required water supplied at the top (**headrace**) and carried away at the bottom (**tailrace**). The water flowed into buckets, and the weight of the water generated torque, which turned the wheel.

- The overshot wheel was far more efficient than the undershot wheel. In the undershot wheel, the water always moved faster than the vanes. In the overshot wheel, the water was captured in the buckets; thus, its full weight contributed to the wheel's rotation.
    - Because of this efficiency, a typical Roman overshot wheel could generate at least 2 horsepower: 30 times better than an undershot wheel of similar size.
    - Further, it could operate with a surprisingly low quantity of flow. The stream had to supply only enough water to fill the buckets continuously.

- The principal disadvantage of the overshot wheel configuration was the need to supply water at the top of the wheel and then remove it at the bottom. For a typical overshot wheel located along a river or stream, the headrace might have to run far upstream and the tailrace far downstream to achieve the necessary change in elevation. Construction of the channels, along with the need for watertight buckets, would have made the overshot wheel much more expensive than an equivalent undershot wheel.

- For this reason, the undershot wheel was probably the preferred technology in most practical circumstances, despite its low power output. At sites where the water supply was adequate, efficiency would not be an issue as long as the wheel generated enough torque to turn a millstone. The overshot wheel would have been used primarily at locations where the water supply was too limited to drive an undershot wheel or where water was already available above the wheel.

Right-Angle Gearing

- Both the overshot and undershot configurations required right-angle gearing. Like so many other machines we've discussed, gear systems were probably developed in Alexandria in the 3rd century B.C. Gears were wooden disks with pegs around the circumference. Vitruvius called these gears ***dentatum***—or "toothed disks." In a geared milling system, the spindle was supported on a horizontal beam (**bridge tree**), hinged at one end and raised or lowered to adjust the millstone.

- In general, gearing is used for three purposes: to change the direction of a rotating shaft, to increase the speed of rotation, or to increase torque. All gearing systems in Roman mills were used for the first purpose—to change horizontal shaft power into vertical shaft power. But the other two purposes are mutually exclusive: To increase speed, a larger gear drives a smaller one; to increase torque, a smaller drives a larger.

- Later overshot wheels had much more power and, thus, were probably geared for increased speed to achieve higher productivity. To accomplish this, the *dentatum* driving the millstone was replaced with a lantern pinion, composed of two wooden disks connected by iron rods. The pinion's diameter was much smaller than that of the driving gear, which means this arrangement would indeed have increased speed at the expense of torque.

- From the available evidence, it is reasonable to conclude that there was no single standard configuration of Roman mill gearing. Rather, it seems that Roman engineers understood the tradeoff between torque and speed and adapted the gear ratio of each mill installation to meet the unique circumstances at hand.

The Vertical-Shaft Wheel

- The third type of waterwheel is the less common vertical-shaft wheel. Typically, the wheel was mounted at the bottom of a pit, oriented horizontally, and drove a vertical shaft. A steeply sloped headrace was used to direct a high-velocity stream of water tangentially onto the angled blades of the wheel, and then the water was drained into a tailrace at the bottom of the pit. As the wheel rotated, the shaft turned a millstone at its upper end.

- The vertical-shaft wheel is often claimed to be the oldest form of water-power generation—in part, because Hero of Alexandria attested to its existence in the 3rd century B.C., but also because it was the simplest of the three configurations. The wheel shaft drove the millstone directly, without the complicated gearing system that the overshot and undershot wheels required.

- o Yet this advantage of the vertical-shaft wheel was also its greatest disadvantage. Because the system used no gearing, the millstone was constrained to rotate at the same speed as the wheel—which eventually was regarded as too slow for efficient milling.

- o This may explain why vertical-shaft wheels were not as common as the other two types.

- Ironically, this oldest of all power generation technologies is amazingly similar to the water turbines used in today's state-of-the-art hydroelectric power plants. Indeed, a 4th-century Roman mill at Chemtou, Tunisia—the oldest surviving vertical-shaft installation—is widely regarded as the world's first true turbine engine. This sophisticated machine is a powerful reminder that many technologies we think of as modern are actually products of ancient technology.

Important Terms

bridge tree: A horizontal beam that supports the vertical shaft of a mill and allows adjustment of the upper millstone's height.

catillus: The hourglass-shaped upper stone in a rotary mill; the *catillus* rotates on the lower stone, or *meta*.

dentatum ("toothed disk"): Roman term for a gear.

headrace: An artificial channel that delivers water to a waterwheel.

hopper mill: A simple milling device that uses a lever to move the upper stone across the lower one.

meta: The cone-shaped lower stone in a rotary mill; the upper stone, or *catillus*, rotates on the *meta*.

overshot wheel: A waterwheel that is mounted on a horizontal shaft and driven by water flowing into buckets at the top of the wheel.

Pompeiian mill: An animal-powered rotary mill (also called a donkey mill).

right-angle gearing: An assembly of two gears that converts horizontal shaft power to vertical shaft power in a mill.

rotary quern: A milling device consisting of a dome-shaped lower stone, a concave upper stone with a hopper carved into its center, and a wood or iron pivot (called the spindle) on which the upper stone rotated.

saddle quern: A simple milling device consisting of two shaped stones.

spindle: The iron pivot on which a millstone rotates.

tailrace: An artificial channel that carries water away from a water wheel.

undershot wheel: A waterwheel that is mounted on a horizontal shaft and driven by water passing underneath the wheel.

vertical-shaft wheel: A waterwheel that is mounted on a vertical shaft and is driven by water striking the wheel's angled vanes at high velocity.

Suggested Reading

Landels, *Engineering in the Ancient World*, chapter 1.

Oleson, *The Oxford Handbook of Engineering and Technology*, chapters 13–14.

Wilson, "Machines, Power and the Ancient Economy."

Questions to Consider

1. Why was water power used almost exclusively for milling grain during most of classical antiquity?

2. Why did a wide variety of other water power applications only emerge in the late Roman Empire?

3. Why is an overshot waterwheel so much more mechanically efficient than an undershot wheel?

Machines at War—Siege Towers and Rams

Lecture 20

Wars in antiquity were particularly brutal, often ending with the mass slaughter, deportation, or enslavement of entire populations. In this lecture, we'll examine the darker side of ancient technology: its use in warfare. Many historical events that have profoundly altered the course of human affairs were significantly influenced by military technology. What's more, the urgency of war tends to concentrate funding, creativity, and expertise on technological development—with an intensity rarely seen in peacetime. The technology of war is worth studying because it is an essential part of the broader story of technological development.

Technologically Intensive Warfare: The Siege

- The ancient world's most technologically intensive form of warfare was the **siege**, which provided a powerful stimulus for the development of large-scale machines that were as ingenious as they were deadly.

- Siege warfare probably originated in Mesopotamia during the 2nd millennium B.C. in response to the development of increasingly effective fortifications. To overcome these formidable defensive systems, attacking forces would use any of five different siege methods:
    - Cross over a city's defensive wall, using scaling ladders, mobile siege towers, or earth embankments.
    - Break through the wall with battering rams or undermine the foundations.
    - Tunnel under the wall.
    - Blockade the city and starve it into submission.
    - Use deception or treachery.

Assyrian reliefs show troops scaling city walls using ladders and ramps, wheeled battering rams, and mobile siege towers that served as firing platforms for archers.

- None of these methods guaranteed success. Thus, the ancient world's greatest masters of siegecraft—the Assyrians, Persians, Macedonians, and Romans—learned to employ multiple methods and technologies in flexible and complementary ways to enhance their prospects for victory.

Periteichismos

- Herodotus tells us—and archeology confirms—that the Persians under Cyrus the Great employed the full panoply of siege methods in their conquest of Lydia and the Greek cities of Ionia in the 6th century B.C. Through these campaigns, the Greeks probably got their first taste of siege warfare.

- Despite this, Classical-era Greeks seldom practiced siegecraft themselves. The typical polis simply could not afford to develop siege machinery. Rather, Greek city-states fielded armies of citizen soldiers called **hoplites**—infantrymen who bought their own armor and fought in a tightly packed linear formation called the **phalanx**. Greek hoplites were renowned for their effectiveness on the field of battle, but they were of little use in a siege.

- Pericles, the great Athenian statesman and general, was reported to have been the first Greek to use a battering ram—at a siege of Samos in 440 B.C. This use of siege machinery was a rare exception for the Greeks.

- Even affluent Athens at the height of its imperial power relied almost exclusively on the blockade to coerce recalcitrant subjects back into the Delian League, a group of city-states allied for mutual defense against the Persians. The blockade—called ***periteichismos*** in Greek—was executed by encircling an enemy city with a substantial wall of stone or wood, augmented by a naval blockade, if necessary.

A Think Tank for Siege Warfare

- Large-scale mechanized siege warfare was finally brought to the Western world in the 5th century B.C. by the Carthaginians. Carthage had increasingly come into conflict with the Greek cities of Sicily.

- Alarmed by the Carthaginian onslaught, Dionysius, the tyrant of Syracuse, responded with an extraordinary initiative. He recruited scientists and engineers from all over the Mediterranean world, paid them handsomely, and put them to work designing the world's most advanced siege machines. In effect, he created the first government-sponsored research and development laboratory. By 399 B.C., this "think tank" had produced battering rams, siege towers, and early prototypes of a weapon that would revolutionize warfare: the catapult.

- Armed with these new machines, Dionysius experienced considerable success against the Carthaginians. Yet despite these prominent successes, mechanized siege warfare failed to take hold in mainland Greece for another five decades. It seems that the Homeric ideal of the warrior hero was too deeply embedded in Greek culture to be easily replaced by technology.

Alexander the Great

- When change finally arrived, it came not from the Greek heartland, but from the kingdom of Macedon—where Philip II and his son Alexander (later known as Alexander the Great) were reinventing warfare in their pursuit of world domination.

- Today, both Philip and Alexander are known for their development of a revolutionary battlefield formation that integrated cavalry with a new type of phalanx, but both men were equally adept at siegecraft. Alexander, in particular, demonstrated great vigor and flexibility in his operations against walled cities.

- Alexander's most spectacular siege was conducted in 332 B.C. against Tyre—a Persian-controlled island city. He approached the island by employing tens of thousands of men to build a 1/2-mile-long earthen causeway, over which he advanced two immense siege towers against the walls of Tyre. Alexander also equipped some of his ships with battering rams and used them to conduct a coordinated assault around the city's perimeter wall.

The "City Taker" Siege Tower

- A typical siege tower of this era was the ***helepolis***, or "city taker." It was built for Alexander around 330 B.C. by an engineer named Posidonius.

- The machine was enormous—60 feet long, 50 feet wide, and more than 100 feet tall. It rolled on solid 9-foot-diameter wooden wheels. Inside were protected staircases for the assault troops; near the top was a drawbridge that could span from the tower to the enemy battlements. On top of the tower was a platform, from which archers and catapults could engage enemy defenders. The timber exterior walls were covered with plaster and animal hides for fireproofing and to cushion the shock of enemy missile impacts.

- Recognizing that the *helepolis* could move only on a flat, hard surface, cities soon began supplementing their fortifications with obstacles along potential avenues of approach. Alexander's engineers then responded with the ditch-filling **tortoise**, a four-wheeled armored shelter for work crews preparing the way for the *helepolis*. The outer shell of this vehicle was padded and fireproofed, and its walls were sloped to deflect enemy projectiles thrown from above. An ingenious chassis configuration allowed the wheels to pivot 90 degrees.

Hegetor's Ram Tortoise: More Impressive Than Effective

- In the Hellenistic era that followed Alexander's death, siegecraft became more important than ever, as the successor kings fought for control of the greater Greek world. During this period, Hellenistic armies built ever-larger and ever-more-impressive siege towers and rams. A representative example of these Hellenistic monstrosities was the ram tortoise of Hegetor, named for the Byzantine engineer to whom it is credited.

- According to ancient sources, Hegetor's ram tortoise had an armored pyramidal shell, measuring 60 by 40 feet and running on eight 6-foot-diameter wheels. Above its timber chassis was an intermediate floor that mounted several arrow-shooting catapults, and above that was a 12-foot-tall turret. The iron-headed ramming beam was more than 100 feet long and wrapped in rawhide for fire protection.

- As impressive looking as this machine must have been, the ram tortoise of Hegetor could have been only marginally effective in battle. When a ramming beam is suspended from above, the amount of energy available to batter an enemy fortification is a function of just three factors—the weight of the ramming beam, the force applied to pull it back for each blow, and the length of the suspension ropes on which the ramming beam swings. Greater destructive power can be achieved with a heavier beam, greater pulling force, and longer suspension ropes.

The Roman Approach to Siegecraft

- The trend toward ever-larger siege machines ended as the Hellenistic kingdoms yielded to the ancient world's newest masters of siegecraft: the Roman legions. The Roman approach to siegecraft emphasized efficiency and functionality in lieu of the Hellenistic-era emphasis on grandiose scale and imposing appearance. Unsurprisingly, the Roman approach proved far more effective.

- Like the Greeks, the Romans were slow to adopt mechanized siege methods. Initially, they favored the storming assault—supported by scaling ladders and little else. But starting around 200 B.C., the legions adopted the ramped embankment, or ***agger***, as a characteristic feature of Roman siegecraft.

- The *agger* was constructed of soil and rubble and reinforced with timber. To protect its builders, the Romans developed portable shelters: hide-covered wooden frames called ***vinea***. These modules could be linked together to create a protected corridor, through which workmen could carry baskets of rubble forward in comparative safety.

- Initially, the *agger* was intended simply as an improvement over the scaling ladder—a means of facilitating massed infantry assaults on enemy battlements. But during the 2nd century B.C., Roman military engineers began using it as a ramp for battering rams and siege towers.
  - Machines operating from an elevated position atop the *agger* could be much smaller than their Hellenistic predecessors, which operated from ground level.

  - This idea led to the development of smaller, more mobile Roman war machines, which were standardized and incorporated into a methodical, flexible system of siege warfare.

The Roman Ram Tortoise: Lighter and More Maneuverable

- The Roman siege of Jotapata in A.D. 67, during the Jewish-Roman Wars, exemplifies the Roman approach to siegecraft. Led by the future emperor Vespasian, the operation began with a series of Roman infantry assaults, which were vigorously repulsed by Jewish defenders under the command of Joseph ben Matityahu (Josephus). Vespasian then directed the construction of an *agger*.

- Roman troops systematically denuded the local countryside of stone and timber and then built the *agger* under covering fire from 160 catapults. Yet the defenders thwarted this effort by raising the height of their wall at the terminus of the ramp.

- Exasperated, Vespasian changed his tactics—deciding instead to blockade Jotapata and starve its defenders into submission. This approach might have worked because the city was desperately short of water, but Joseph thwarted Vespasian once again by resorting to psychological warfare. He had his men soak their outer garments in water and hang them from the battlements. The deception worked. Convinced that the Jews must have an abundant supply of water, Vespasian gave up the blockade and ordered a renewed assault.

- This time, the Romans deployed a ram tortoise up the *agger*, again supported by catapult fire.
  - Unlike the Greek counterpart, the standard Roman ram tortoise was small and could be moved easily by 20 men. Its structure was simple yet strong. It was covered with 3-inch-thick planking and fireproofed with clay mixed with hair. Its best defensive feature were steeply sloped sides to deflect projectiles.

  - Ultimately, the ram succeeded in battering down the upper wall, and Vespasian's men charged into the breach, locking their shields in the ***testudo***, or "tortoise," formation.

  - Undaunted, the Jews doused the attackers in boiling oil and repulsed them yet again. Vespasian then brought up three 50-

foot siege towers, their outer walls reinforced with iron plates. From these towers, catapults provided accurate covering fire, while engineers raised the *agger* even higher.

- Finally, on the 47th day of the siege, the embankment topped Jotapata's upper wall—and that night, Vespasian's son Titus led a small force over the battlements. They surprised the Jewish sentries and threw open the gates for the entire Roman force, which easily captured the city. The siege ended as most ancient sieges did—with most of Jotapata's intrepid defenders killed and its women and children sold into slavery.

Important Terms

agger: A ramped embankment used in Roman siege warfare.

helepolis ("city taker"): A mobile siege tower.

hoplite: A heavily armored Greek infantryman, equipped for fighting in the phalanx formation.

periteichismos: A Greek siege tactic, in which the besieging force surrounded and isolated the besieged city with a substantial wall of wood or stone.

phalanx: A battlefield formation consisting of multiple ranks of closely spaced, heavily armored infantry called hoplites.

siege: A form of warfare aimed at overcoming fortified defensive positions. Siege warfare often employed heavy machinery, such as mobile towers, battering rams, and catapults.

testudo ("tortoise"): A Roman infantry formation used for assaulting fortified positions. Soldiers formed a testudo by locking their shields together over their heads to form a protective barrier.

tortoise: An armored wheeled vehicle used in siege warfare.

vinea: Hide-covered wooden frames linked together to create a protected corridor, through which workmen and assault troops could approach a fortified position in comparative safety.

Suggested Reading

Campbell, *Besieged.*

———, *Greek and Roman Siege Machinery, 399 BC–AD 363.*

Oleson, *The Oxford Handbook of Engineering and Technology*, chapters 26–27.

Questions to Consider

1. To what extent do the various forms of siege warfare (Assyrian, Persian, Classical Greek, Macedonian, Hellenistic, Roman) reflect the political and social contexts from which they emerged?

2. Compare and contrast the Hellenistic ram tortoise of Hegetor with the Roman ram tortoise designed by Apollodorus of Damascus. Which was the more impressive technological achievement? Which was the more effective siege weapon?

3. How did Classical Greek and Roman attitudes toward the use of technology in warfare differ?

Machines at War—Evolution of the Catapult

Lecture 21

Of the many weapon systems in ancient history, none was subject to more assiduous development and thorough documentation than the catapult. For these reasons, the catapult serves as an instructive 700-year case study for the process of technological development in the Classical world. In 399 B.C., under Dionysius of Syracuse, a team of scientists and engineers took on a number of technological challenges to design and build state-of-the-art weapons. The challenge with the most far-reaching consequences was their attempt to overcome the inherent human physiological limitations associated with the conventional handheld bow and arrow. Their efforts eventually led to the transformation of the bow-and-arrow system into something fundamentally new: the catapult.

Improving the Bow and Arrow

- One of the first machines of war (and other activities, such as hunting) was the bow and arrow. As a technological system, the bow and arrow depends largely on the bow's ability to store **elastic energy**. More elastic energy in the bow translates to more **kinetic energy** imparted to the arrow, resulting in higher arrow velocity and, therefore, longer range.

- A particularly useful way to characterize the performance of a bow is to plot a graph of pulling force versus draw length—also called a **force-draw curve**. In order to improve the performance of the bow, we must change its physical characteristics in ways that increase the elastic energy. There are three ways to accomplish this:
  - Increase the length of the draw and, thus, increase the stored energy.

  - Pre-tension the bowstring by shortening it; this imparts some initial curvature to the bow, which gives a "head start" on the draw and further increases stored energy.

- Make the bow stiffer—either by making it thicker or by using a different material. A stiffer bow requires more force to bend, resulting in significantly more elastic energy.

- As early as the 2nd millennium B.C., bow makers discovered that they could achieve even better performance by building the bow from three different materials: wood, animal horn, and animal tendon (or sinew). Thus was born the **composite bow**—one of the most ingenious applications of biological materials in human history.

The *Gastraphetes*, or "Belly Bow"

- In the Classical era, the performance of the composite bow was limited, not by the mechanical characteristics of the bow, but by the physiological limitations of the archer. Because of these limitations, the optimal bow was one that required 45 pounds to achieve a maximum draw of 30 inches. Significantly more powerful composite bows were quite feasible—but there was no point in building them because human archers could not take advantage of the enhanced capability.

- Dionysius's "think tank" on weapon development sought to solve this problem by developing the ***gastraphetes***, or "belly bow." The *gastraphetes* consisted of a powerful composite bow attached to a wooden stock. A wooden slider was dovetailed into the stock, so it could slide forward and back. Its upper surface had a groove to guide the arrow; it also had an iron trigger mechanism mounted on its back edge. Because of these enhancements, the *gastraphetes* could use a larger, more powerful bow than its hand-drawn predecessors. Thus, it had significantly longer range and could shoot heavier arrows.

- The *gastraphetes* was used for the first time in 397 B.C. at Dionysius's siege of Motya—where it apparently made an impression on the Carthaginians. It triggered an arms race that only abated in the waning years of the Roman Empire.

Evolution of the Catapult

- Seeking further improvements in performance, the builders of later *gastraphetes* used progressively larger and more powerful bows. Eventually, a windlass was mounted on the rear of the stock and used to pull the slider back. This device produced significant mechanical advantage, but it also made the weapon heavier and more unwieldy. As a result, the stock had to be mounted on a fixed base, which facilitated handling and significantly improved the weapon's accuracy.

- At this point, we are seeing something fundamentally new—a true catapult: an artillery weapon capable of shooting large iron-tipped arrows (called bolts) or 50-pound stone balls. This stage of development was exemplified by the stone-throwing catapult built by Charon of Magnesia in the 4th century B.C.

- Impressive as it must have been, however, Charon's machine proved to be an evolutionary dead end. The composite bow had finally reached its technological limit. Further improvements in catapult performance could be achieved only by devising a fundamentally new means of storing elastic energy. This need gave rise to the quintessential artillery weapon of the Hellenistic and Roman eras: the **torsion catapult**.

The Torsion Catapult

- Development of the torsion catapult occurred in a series of small increments, starting in the mid-4th century B.C. By the end of the century, this process had produced two standardized machines: an arrow shooter called the **euthytone** and a stone thrower called the **palintone**. Central to the palintone was something fundamentally new: a pair of vertically oriented **torsion springs**, which replaced the bow as the means of storing elastic energy.

- **Torsion** is the twisting of a structural element about its axis. In the palintone, each torsion spring was a bundle of rope with a wooden arm thrust through its center. When the weapon was cocked, these arms rotated rearward, and their rotation was resisted by the twisting of the springs. As the springs twisted, they stored elastic energy.

- Most Classical-era catapult springs were made of sinew—the same material that was used so effectively in the composite bow. Because of its extraordinary ability to store elastic energy, sinew was the ideal material for catapult springs.

- When the trigger was pulled, some of the energy stored in the torsion springs was used to accelerate the projectile, but a significant portion was also used to accelerate the wooden throwing arms. This latter quantity of energy would be wasted, except that the forward motion of the arms was eventually arrested by the sling snapping taut.

- Because this tightening of the sling also contributed to the forward motion of the projectile, nearly 100 percent of the energy released from the palintone's torsion springs was ultimately transmitted into the projectile as kinetic energy—a truly extraordinary degree of mechanical efficiency.

Early Technology Transfer

- Although the catapult's effect on the conduct of warfare was substantial, its influence on math, science, and engineering was perhaps even greater.
    - Sometime around 275 B.C., scientists and engineers at the Museum of Alexandria began developing a rigorous empirical design methodology for the torsion catapult.

    - As documented by both Vitruvius and a Greek engineer named Philon, this methodology consisted of two steps: First, an experimentally derived formula was used to calculate the diameter of the torsion spring. Second, this dimension was used as a module for determining the size of every other component in the machine.

- For the palintone, the formula for the torsion spring diameter was $D = 1.1\sqrt[3]{100\,M}$, where D is the diameter, measured in ***dactyls*** (a

Greek unit of length, equal to about ¾ inch) and M is the projectile weight in Attic ***minas*** (a unit of weight, equal to about 1 pound). It was, amazingly, scientifically correct.

- o Today, we know that the elastic energy stored in a torsion spring is a function of the volume of the spring, and the kinetic energy of the projectile is a function of its weight; therefore, the mathematically optimum spring diameter should be a function of the cube root of the projectile's weight—just as the formula indicates.

- o The fact that this empirical formula correctly reflects the underlying scientific principles is a tribute to the rigor of the experimental process that was used to derive it.

- Perhaps most astonishingly, when this formula was first developed, no mathematician had yet figured out how to calculate a cube root. It was not until later that the Greek mathematician Eratosthenes devised a sophisticated graphical method for calculating the cube root of a number and invented a mechanical device to implement this method. Eratosthenes was quite explicit about his reason for working on the cube root problem in the first place: He wanted to facilitate catapult design.

- Greek scholars used rigorous experimentation to derive a scientifically correct mathematical model to facilitate optimal engineering design. They then devised a mechanical tool to perform advanced computations in support of that process. This was a stunning example of the integration of math, science, and engineering—entirely modern in character yet entirely unique in classical antiquity. And it was all about catapults.

Later Developments of the Catapult

- In modern engineering practice, the availability of powerful mathematical, scientific, and computational tools often serves as a stimulus for innovation. If engineers can accurately predict how various design alternatives will perform without having to build and test a prototype for each one, then they are more likely to explore a wider range of alternatives—and to arrive at an innovative solution. After the mathematical model for catapult design was formulated around 275 B.C., a variety of new developments in catapult technology emerged. Consider a few examples of these later developments:

- A series of experimental catapults—attributed to the Greek inventor Ctesibius—replaced the torsion springs with bronze leaf springs and pneumatic cylinders.

- The **scorpion**, a Roman arrow shooter, achieved greater power by using forward-curving arms.

- The ***cheiroballistra***—a small, mobile Roman arrow shooter—used iron rather than wood for its main frame and added bronze cylinders to protect the torsion springs.

- The Hatra ballista, a revolutionary Roman machine with inward-swinging arms, yielded substantial improvements in power and range because its arms could rotate through a significantly larger angle than those of a standard palintone.

- The one-armed **onager**—a late-Roman stone thrower—was far less mechanically efficient than a two-armed machine but was easier to build and maintain.

- Clearly, the catapult was far more than just a weapon of war. For seven centuries, it served as a mirror, reflecting the many faces of technological development in the Classical world. More than any other engineered system, the catapult speaks to us about the cultural and political world from which it emerged.

Important Terms

cheiroballistra: A small, mobile Roman arrow-shooting catapult, which used iron rather than wood for its main frame.

composite bow: A bow constructed from three different materials—animal sinew (in tension) and bone (in compression), around a wooden core.

dactyl: A Greek unit of measurement, equal to approximately ¾ inch.

elastic energy: Energy stored in the deformation of a material—e.g., the bending of a bow (also called strain energy).

euthytone: An arrow-shooting torsion catapult.

force-draw curve: A graph of the force required to draw a bow versus the distance drawn.

gastraphetes ("belly bow"): An arrow-shooting forerunner of the catapult. The *gastraphetes* incorporated a stiff bow and a mechanical configuration that allowed the archer to cock the weapon by leaning his full weight on the stock.

kinetic energy: The energy associated with a mass in motion.

mina: A Greek unit of weight, equal to approximately 1 pound.

onager: A late-Roman catapult with only one throwing arm.

palintone: A stone-shooting torsion catapult (also called a ballista by the Romans).

scorpion: A Roman arrow-shooting catapult with forward-curving arms.

torsion: Twisting of a structural or mechanical element.

torsion catapult: A weapon that used two vertically oriented torsion springs and two throwing arms to shoot heavy arrows (called bolts) or stones.

torsion spring: A spring consisting of a bundle of rope made from animal sinew. Elastic energy is stored in the spring by twisting it.

Suggested Reading

Campbell, *Greek and Roman Artillery, 399 BC–AD 363*.

Marsden, *Greek and Roman Artillery: Historical Development*.

———, *Greek and Roman Artillery: Technical Treatises*.

Oleson, *The Oxford Handbook of Engineering and Technology*, chapter 13.

Questions to Consider

1. Why did the catapult serve as such a uniquely powerful stimulus for mathematical, scientific, and technological development in the ancient world?

2. In what ways did the catapult overcome the inherent physiological limitations of a human archer?

3. How did the development of the torsion spring overcome a technological "dead end" that otherwise would have severely limited catapult performance?

Machines at Sea—Ancient Ships

Lecture 22

Plato once observed that the Greek people were like frogs sitting around a pond—their cities hugging the coastline, their attention focused on the water. The ancient Greek world was inextricably linked to the sea. Many poleis were port cities, engaging in seaborne commerce throughout the Mediterranean and often exercising military power by sea, as well. The Romans, by contrast, were not a seafaring people. Romans built capable ships when they had to, but they never did it with the same ingenuity or grand ambition that characterized their contributions in water supply or building construction. In the next two lectures, we'll look at the nautical technologies—the sailing vessels and oared warships—that epitomized the Greek seafaring culture.

The Kyrenia ship is a 4th-century-B.C. merchant vessel; at the time it sank, it was carrying a cargo of wine and millstones.

Roundship and Longship

- In the 3rd millennium B.C., Bronze Age Greeks and Minoans developed early prototypes of two basic ship configurations: the **roundship** and the **longship**, which would remain in use throughout the Classical era and beyond.

- The roundship was invariably a merchant vessel, powered primarily by sail (sometimes fitted with supplemental oars). As the name suggests, it had a broad, rounded hull, typically with a length-to-width ratio of about 4 to 1, to maximize its cargo capacity.

- The longship was generally either a warship or a pirate vessel. Its primary source of power was a large crew of oarsmen, though most longships also carried at least one supplemental mast and sail. Its hull was built for speed, with a length-to-width ratio as high as 10 to 1. There were many types and sizes of longships, but nearly all had a characteristic profile: long and low, with a ram extending forward from the bow at the waterline and a dramatic, upward-curving sternpost.

Shell-First Construction

- Ships built after the Classical era, such as the U.S.S. *Constitution*, were constructed using the **plank-on-frame** method. This began with assembly of a heavy timber framework consisting of a **keel**, stem post, sternpost, and a series of closely spaced wooden frames. This rigid framework established the shape of the hull and served as its structural skeleton. Only after this internal skeleton had been completed was the outer planking added, with the individual planks attached to the frames but not to each other.

- In contrast, all Classical-era hulls—roundship and longship—were constructed using the **shell-first** method. The entire outer shell of planking was built first and then interior frames were added subsequently. To create the shell, the individual planks were attached to each other along their edges. Many shipwrecks of shell-first hulls showed no evidence of caulking between the planks—a good indicator that this method produced extremely tight-fitting joints.

- The principal source of structural strength in a shell-first hull was the shell itself. The shell-first method produced a hull that was amazingly strong and light; however, the process was also quite labor intensive. Even a small ship would have had many thousands of **mortise-and-tenon joints**, and every one had to be fitted precisely to ensure the structural integrity of the hull.

The Concept of Buoyancy

- According to the concept of **buoyancy**, any floating object displaces water. The weight of the displaced water equals the weight of the floating object. This displacement of water causes an upward force, called the buoyant force, which exactly counterbalances the boat's weight and causes it to float.

- When a ship floats in still water, the buoyant force is uniformly distributed along its length; thus, it balances the ship's weight, and the hull experiences relatively little tendency to bend. But in rough seas, the buoyant force can be distributed quite unevenly along the length of the hull; when this happens, the hull bends like a beam. There are two worst-case conditions:
  - **Hogging**—When the crest of a wave is amidships, the buoyant force is concentrated in the middle, and the entire hull bends concave downward.

  - **Sagging**—When the trough of a wave is amidships, the buoyant force is concentrated at the bow and stern, and the hull bends concave upward.

Ancient Joinery

- When a ship hull bends, planks slide horizontally across each other, which is called shearing. In the hull of a Classical-era ship, this critical structural function—preventing shearing along the seams between hull planks—was accomplished by thousands of mortise-and-tenon joints.

- To resist bending, a hull also depended heavily on its keel: the ship's structural backbone. For longships, in particular, it was generally necessary to assemble the keel from two or more lengths of timber, joined together with splices. This was a significant structural challenge, because these joints had to be capable of carrying the same extreme loads as the keel itself. In ancient times, the wooden splice had to do the job on its own.

- To meet this challenge, ancient shipwrights devised an amazing type of joinery. The **locking scarf joint** has been found in many ancient shipwrecks and appears to have been the standard method of splicing keels in Classical-era ships.

Harnessing the Wind

- The sailing ship was the ancient world's only practical application of wind power. A typical Greek merchant vessel was propelled by a single square sail, mounted on a mast. The mast was mounted in a socket, called the **step**, down in the ship's keel, and then held in place with ropes called **stays** (which extended fore and aft) and **shrouds** (which extended to port and starboard). All these ropes were called standing rigging because they were not moved.

- The ropes associated with controlling the sail were called running rigging, because they were constantly being adjusted. The sail was mounted on a horizontal **yardarm**, which could be raised and lowered by a double line called the **halyard**. Ropes called **lifts** were used to tilt the yardarm; **braces** were used to rotate it horizontally.

- The sail itself was controlled by lines called **sheets**, which ran from the sail's two lower corners back to the helmsman's position at the stern. Normally, the sheet on the windward side of the ship was made fast, and the one on the leeward side was held by the helmsman, who used it to control the speed of the vessel. When the sheet was pulled taut, the sail caught the full force of the wind. The size of the sail was adjusted using **brails**.

Sailing into the Wind

- The various components of running rigging, combined with two steering oars, provided a flexible system for controlling the ship's direction and speed. Indeed, this system was so effective that it allowed Greek ships to sail into the wind—probably for the first time in nautical history.

- Sailing into the wind is all about the mechanics of forces. Sailing into the wind is nautical terminology: A mariner's compass has 32 **points**; thus, one point is 11¼ degrees. For the sake of comparison, modern sailboats can generally sail two points into the wind—about 22 degrees. The ship makes forward progress by **tacking**: zigzagging back and forth across the wind repeatedly at that 11-degree angle.

- Sailing into the wind with a square-rigged vessel is both challenging and inherently inefficient. Not only must the ship travel up to five times farther than the desired straight-line route because of tacking, but it can harvest only about 40 percent of the wind energy that would be available if it were sailing with a "following wind." What's more, the vessel's forward progress is further hindered by leeward drift.

- Overcoming these challenges required an advanced hull design, combined with flexible and precise systems for steering the ship and controlling the position, size, and shape of the sail. Greek shipbuilders addressed this problem in the 5th century B.C. with an innovative hull design, shaped like a wineglass, that greatly reduced susceptibility to leeward drift.

The All-Important Second Sail

- Wind and sail mechanics can shed light on another important Classical-era development in nautical technology. When a ship sails across the wind, it heels over to the leeward side. This sideways lean shifts the line of the wind force off the hull's centerline, causing a torque that turns the ship into the wind: a tendency that mariners call **weather helm**.

- For a ship with one sail, weather helm could be countered only by the helmsman turning the steering oars in the opposite direction—a corrective action that slowed the ship. Greek shipbuilders eventually realized that they could address this problem more efficiently by adding a small mast and sail near the ship's bow. This foresail could be trimmed to counterbalance the weather helm, while providing some additional propulsion.

- Over time, this insight—that a second sail added flexibility in controlling the ship—led to the development of large three-masted Greek and Roman freighters, which prefigured the grand oceangoing ships of the Age of Discovery more than a millennium later.

Unprecedented Levels of Performance

- Because of their technological refinement, Classical-era roundships achieved unprecedented levels of performance under sail. These ships carried large cargoes over great distances, despite the challenges of Mediterranean weather. An ordinary merchant ship of this era would have been about 60 feet long and 15 to 20 feet wide, with a carrying capacity of 120 to 150 tons.

- However, ships carrying 400 to 500 tons were common, and several carrying more than 1,000 tons are described in the literature. Typical cargoes included grain, wine, building stone, metal ingots, oil, and during the Roman era, looted art treasures. The development of these vessels was a substantial technological achievement—and remains a vivid illustration of Greek nautical engineering genius.

- In the next lecture, we'll examine the ancient longship—a human-powered torpedo designed and optimized for the sole purpose of victory in war.

Important Terms

braces: A pair of ropes extending from the tip of a yardarm to the ship's stern; used to rotate the yardarm horizontally.

brails: Light ropes attached to the lower edge of the sail that extend upward over the yardarm and then back to the stern, where they are tied or held by a crewman; used to adjust the size and shape of the sail.

buoyancy: The scientific principle governing floating objects. When an object floats, the weight of the water it displaces is equal to the weight of the object.

halyard: A pair of ropes from which the yardarm is suspended.

hogging: The bending of a ship's hull caused by a large wave amidships. A hogging hull bends concave downward.

keel: The structural backbone of a ship.

lifts: A pair of ropes extending from the tip of a yardarm to the top of the mast; used to support the yardarm and tilt it vertically.

locking scarf joint: An elaborate joint used to splice two pieces of wood together for a ship's keel.

longship: A ship with a long, narrow hull, powered primarily by rowers and generally used as a warship or pirate vessel.

mortise-and-tenon joint: A joint used to fasten planks to each other in a shell-first hull.

plank-on-frame: Construction method used in most wooden ships after the Classical era. The plank-on-frame hull was created by assembling the keel, stem post, sternpost, and wooden frames into a rigid framework. Planking was added only after this framework was complete. Planks were attached to the frames but not to each other.

point: An angle equal to 1/32 of a full circle (or 11¼ degrees).

roundship: A ship with a broad, rounded hull, powered primarily by sail and generally used for carrying cargo.

sagging: The bending of a ship's hull caused by large waves fore and aft and a trough amidships. A sagging hull bends concave upward.

sheet: A pair of ropes extending from the lower corners of the sail to the helmsman's position at the stern of the ship; used to control the sail.

shell-first: Construction method used for most ancient ship hulls. The shell-first hull was created by attaching individual planks to each other along their edges, using mortise-and-tenon joints. Light wooden frames were added to the inside of the hull only after it was completely formed.

shrouds: Ropes that support a ship's mast in the port and starboard directions.

stays: Ropes that support a ship's mast in the fore and aft directions.

step: The socket in a ship's keel, used to support the mast.

tacking: A sailing maneuver in which the ship turns its bow through the wind in order to receive the wind on its opposite side. By executing this maneuver repeatedly, the vessel can make progress directly into the wind, albeit along a zigzag path.

weather helm: The tendency of a ship sailing across the wind to turn into the wind.

yardarm: A horizontal spar from which a sail is suspended.

Suggested Reading

Casson, *Ships and Seamanship in the Ancient World.*

Landels, *Engineering in the Ancient World*, chapter 6.

Oleson, *The Oxford Handbook of Engineering and Technology*, chapter 24.

Questions to Consider

1. Why were the Greeks—and not the Romans—the ancient world's master shipbuilders?

2. How did ancient shell-first hulls achieve such high strength and light weight?

3. How does a sailing ship sail into the wind?

Reconstructing the Greek Trireme

Lecture 23

The trireme is the ancient world's single most fascinating technological system. First, from the perspective of engineering sophistication, longship development reached its high-water mark unusually early—with the trireme in 5th century B.C.—and then declined thereafter. Second, although we have some archeological remains from most forms of ancient technology, not a single remnant of a trireme has survived from antiquity. Third, the trireme is the only form of ancient technology for which engineering optimization was an absolute, nonnegotiable design criterion. Its effectiveness depended on a perfectly optimized balance of speed, maneuverability, and strength. In a sea battle, achieving this perfect balance was, literally, a matter of life or death.

The Penteconter, a 50-Oared Bireme

- Primitive multi-oared longships first emerged in the eastern Mediterranean during the 3rd millennium B.C. By the late 2nd millennium—the era of the Trojan War—the Greek longship had evolved into a true seagoing vessel, propelled by a single bank of oars, which were rowed over the top of the **gunwale**. Homer's *Iliad* describes the Achaean heroes rowing to Troy in a 100-foot-long, 50-oared vessel called the **penteconter**.

- Four centuries after the Trojan War, a new technology appeared on the scene: the **bireme**, a warship with oars on two levels. Adding that second bank of oars was intended, not to double the number of rowers, but to reduce the penteconter's length by half. Shortening the hull made it both stronger and significantly more maneuverable. This development most likely marked a major change in the longship's primary role—from troop transport to offensive weapon.

- With its 50 rowers arranged on two levels, this shorter, stronger penteconter would be better able to outmaneuver and ram enemy ships, without compromising the human power available to propel it. This major change in naval tactics was confirmed by contemporary graphical representations of longships, which for the first time, clearly showed bronze rams affixed to their bows.

- The bireme represented a significant nautical engineering challenge. Two levels of oars required a taller hull, which would have been more unstable. Although stability could be enhanced by adding ballast within the ship's hold, this extra weight would have reduced speed and maneuverability. To minimize height without adding weight, bireme builders developed an optimized configuration with the upper-level oarsmen rowing over the gunwale and the lower-level oarsmen positioned in the hull, rowing through oar ports just above the waterline.

Evolution of the Trireme

- Within just a few decades of the bireme's introduction, ships with a third level of oars began to appear. Thus was born the trireme: the ultimate manifestation of the 8th-century-B.C. naval arms race.

- By the 5th century B.C., the trireme had become the principal means of asserting political power in the eastern Mediterranean. Archeologist John Hale makes a compelling case that the great cultural achievements of Athens' Golden Age would never have occurred without the Athenian naval superiority that derived from the trireme.

- During the reign of Alexander, naval tactics changed yet again, from an emphasis on ramming to an emphasis on boarding and capturing enemy ships. From this point forward, warships became progressively larger and carried significantly larger complements of soldiers—and even artillery.

- First came the quadrireme, which had two banks of oars with two rowers per oar. The quinquereme followed, which most likely had three banks of oars, with two rowers per oar on the upper two levels and one per oar on the lowest level. By the mid-4th century B.C., sixers were commonplace.

- The practical-minded Romans settled on simple triremes and quadriremes as their standard warships. Indeed, Roman naval technological development was focused, not on the ships themselves, but rather, on specialized equipment that would make them more effective as fighting platforms for marine infantry. Characteristic of these developments was the corvus, a hinged boarding ramp that could be lowered across the gap between the Roman ship and its adversary, then used by assault troops as a seaborne version of the *agger*.

The *Olympias* Project

- None of these Hellenistic or Roman ships could compare with the Hellenic-era trireme in speed, maneuverability, and engineering efficiency. Unfortunately, we have surprisingly little substantive evidence for the trireme's technical characteristics. No archeological remains have been found, and the literary and iconographic evidence is both ambiguous and maddeningly incomplete.

- Recently, however, our understanding of this technological system was greatly enriched by an extraordinary exercise in experimental archaeology. In 1982, a team of Classical scholars, naval architects, and rowing enthusiasts initiated a project to design and build a full-scale reconstruction of a 5th-century Athenian trireme.

- The completed vessel was launched in 1987 and christened *Olympias*. This project proved to be an intriguing detective story, requiring the synthesis of Classical scholarship, archeology, expertise in ancient ship construction, and modern naval engineering. It also provides us with some fascinating points of comparison between the engineering of the 5th century B.C. and the engineering of today.

The Trireme Rowers

- Armed with a variety of rich but fragmentary historical pieces of information, the *Olympias* team began designing the trireme's hull. The critical first step was to determine the proper configuration of the rowers—a task that required the team to challenge some conventional wisdom.

- From the Renaissance through much of the 20th century, many scholars maintained that the Greek trireme did not have three levels of rowers. The *Olympias* team was able to demonstrate conclusively that these many generations of scholars were wrong. The Greek trireme could not have used the *alla sensile* (three rowers per bench, with each man pulling his own oar) or the *scaloccio* (three rowers pulling each oar) system and that the trireme did, indeed, have oars on three levels. Here's the key evidence cited by the *Olympias* team:
    - The Athenian naval inventories indicated that most of a trireme's oars were the same size: 9½ cubits (or about 14 feet) long. This was much too short for the *scaloccio* system and lacked the three distinctly different sizes of the *alla sensile* system.

    - An *alla sensile* trireme with 170 oars would have been too long to fit in the Piraeus ship sheds, and a *scaloccio* trireme would be too wide.

    - An Athenian sculptural fragment, the Lenormant Relief, unambiguously depicts a trireme with three banks of oars.

- The team validated this conclusion by demonstrating the feasibility of a three-level oar system using oars of equal length. The top bank of rowers operated from an outrigger projecting beyond the gunwale; the lower-level oar ports were placed very close to the waterline; and the three levels of rowers were staggered longitudinally, so the oars would not interfere with each other during the stroke.

The Trireme Hull

- Having established the rowers' positions, the *Olympias* team worked out the overall size and shape of the hull, ensuring that the following conditions were met:

- The hull could accommodate all 170 rowers in their correct positions and at the correct 2-cubit longitudinal spacing.
- The hull shape was consistent with all known iconographic representations of the trireme, and the stern, in particular, was appropriately configured for hauling the ship onto shore stern-first without incurring structural damage.
- The hull cross-section and construction methods were consistent with those of known contemporary vessels.
- The overall dimensions did not exceed the space provided within the Piraeus ship sheds: 121 feet long and 19 feet wide.
- The hull was positively buoyant and stable without the addition of ballast, and its hydrodynamic properties were consistent with the trireme's known top speed and cruising speed.

- Amazingly, after all feasible options were considered, there was essentially only one hull configuration that met all the historically based characteristics, performance requirements, and constraints.

The Mysterious *Hypozoma*

- When the hull configuration was analyzed as a structural element, a new problem emerged. The long hull was found to be structurally adequate with respect to sagging but severely inadequate with respect to hogging.
- Ancient sources solved the problem. Naval inventories referred to a rather mysterious piece of trireme rigging called the ***hypozoma***, a loop of heavy rope anchored near the ship's bow and stern and used to strengthen the hull with respect to hogging. The *hypozoma* tensioning device used in a trireme was a substantial piece of machinery, indicative of the critical structural function it performed.

The *Olympias* Sea Trial

- During its sea trials, *Olympias* was able to achieve a respectable top speed of 8.9 knots. The ship proved to be highly maneuverable, and it performed quite well under sail. However, its best long-distance cruising speed was only about 6 knots—far short of the cruising speed of 7 to 8 knots attested in the historical sources.

- After much analysis, the researchers arrived at a plausible explanation for this performance shortfall. An average human should be able to exert 1/10 horsepower indefinitely. In the *Olympias*, the rowers were not achieving this level of power output because the oar positions were too close together. Yet Vitruvius had clearly put the distance between rowers as 2 cubits.

- Again, archeology came to the rescue. In the ancient world, such measures as the *dactyl* and the cubit were by no means standardized. And in 1990, archeologists unearthed a relief bearing a slightly longer definition of the cubit. The design team's subsequent analysis showed that a slightly modified version of the *Olympias*, based on this longer cubit, would have the necessary propulsive power to overcome the performance shortfall noted in the sea trials.

- The *Olympias* team only built one ship but used modern scientific tools as an alternative to the Greeks' **empirical design** iterations, confidently predicting how the ship would perform before even starting construction. Therein lies the most fundamental difference between the engineering of antiquity and the engineering of today.

- The *Olympias* project was proclaimed an unequivocal success. This amazing endeavor demonstrated that the trireme was an extraordinarily sophisticated technological system—one that reflected the Greeks' deep understanding of nautical engineering and their superb shipbuilding skills.

Important Terms

bireme: A warship with oars on two levels.

empirical design: The process of designing a structure, machine, or system by making a succession of well-reasoned incremental changes based on experience and the observed performance of previous designs.

gunwale: The heavy wooden rail at the top edge of a ship's hull.

hypozoma: A loop of heavy rope extending from the bow to the stern of a Greek trireme. When tightened by twisting, the ship's two *hypozomata* strengthened the hull against structural failure by hogging.

penteconter: A 50-oared longship.

Suggested Reading

Casson, *Ships and Seamanship in the Ancient World.*

Hale, *Lords of the Sea.*

Landels, *Engineering in the Ancient World*, chapter 6.

Morrison, Coates, and Rankov, *The Athenian Trireme.*

Questions to Consider

1. Why did the Greek trireme evolve into one of the ancient world's most highly optimized technological systems?

2. What methods, materials, and design features did the ancient shipbuilders use to achieve this level of optimization?

3. How does the *Olympias* project inform us about the similarities and differences between the ancient and modern engineering design processes?

The Modern Legacy of Ancient Technology
Lecture 24

This final lecture in our course is about learning to experience the spirit of Classical Greek and Roman technology in our modern world. An excellent example is the war memorial at West Point. The building's exterior is dominated by fluted Ionic columns. The cornice is punctuated by lion's head waterspouts—a wonderful Classical detail. The front entrance, with its pediment and bronze door, is straight out of Book IV of Vitruvius's *De Architectura*. Inside are Corinthian columns and a coffered ceiling like the ones used in Athenian temples. What better way to honor the noble sacrifice of its graduates who died in defense of freedom than to clothe this great building in the trappings of Classical Greece: the first democracy.

Loss of Technologies

- Many of the technological developments of the Classical world were largely lost with the demise of Roman civilization in the 5th century A.D. and had to be reinvented in later eras. No doubt, the fall of the western Roman Empire initiated a "dark age" for technological development and disrupted the well-established system by which engineering expertise was developed, taught, and propagated.

- Nowhere was this phenomenon more apparent than at the Museum of Alexandria, the government-sponsored think tank that inspired so many technological innovations. During its heyday in the 3rd century B.C., the museum employed more than 1,000 of the Hellenistic world's best minds—performing scientific research, inventing, lecturing, and publishing their work. But by the 3rd century A.D., the museum was being actively suppressed and was ultimately destroyed by fire in A.D. 272.

- With the museum's demise, many promising lines of technological development died. Consider the work of Hero of Alexandria, perhaps the museum's most prolific and creative experimenter.

This amazing individual wrote technical treatises on pneumatics, mechanics, surveying, catapult design, optics, and geometry. He is credited with inventing the syringe, vending machine, wind-powered organ, and **aeolipile**—considered to be the world's first steam engine.

- Many other technologies were also lost with the fall of Rome. For example, concrete construction (the hallmark of the Roman construction revolution) ceased to exist, and urban planning methods were largely forgotten. Even the quintessential elements of Classical architecture—the Doric, Ionic, and Corinthian orders—were discarded, as medieval architecture evolved toward the pointed arches, ribbed vaults, and flying buttresses of the Gothic era.

Technological Continuity

- Yet there was also much technological continuity through the period of late antiquity into the Middle Ages. Water power, metallurgy, glassmaking, and other forms of technological know-how survived in isolated pockets throughout the Western world—often preserved and further developed by medieval monastic orders. Thanks to this continuity, we can legitimately claim that modern hydroelectric power plants, steel mills, and fiber optics are indeed legacies of ancient engineering.

- Some forms of construction technology also survived the fall of Rome, largely because of the Roman basilica. This secular building became the predominant model for early Christian churches, primarily because the logical alternative—the Greek-style temple—was ill suited for congregational worship. Today, innumerable small-town churches throughout the Western world continue to use this distinctly Roman form.

- As the basilica endured, so did its principal structural engineering technologies: the arch and the tie-beam truss. The truss has undergone continuous development and is now used in ambitious long-span structures, such as the retractable roof of the University of Phoenix Stadium, and in tall towers, such as the Tokyo Sky Tree.

- Many Roman technological creations were so well built that they survived the fall of Rome and, millennia later, became models for modern builders.
    - For example, in the 17th century, French engineers systematically catalogued and analyzed Europe's many surviving Roman roads, then incorporated their key design features into standards for modern roads.
    - Similarly, surviving Roman aqueducts inspired the design of many modern water supply systems—as is quite evident in the Harlem River crossing of the Croton Aqueduct, a component of the New York City water system, constructed around 1840.

The spectacular Tokyo Sky Tree is the direct descendant of the simple wooden trusses that spanned the Roman basilicas of old.

The Byzantine Empire

- Perhaps the most important source of technological continuity following the fall of the western Roman Empire was the eastern Roman Empire, which evolved into the civilization we call the **Byzantine Empire** and continued to thrive for another 1,000 years.

- Byzantine structural engineering was a direct outgrowth of the Roman imperial building system. The quintessential Byzantine structure was the great church of Hagia Sophia, located in Constantinople (modern Istanbul) and consecrated in A.D. 537. This extraordinary building has been characterized as one of the most ambitious and original in all of human history—yet in its grand dome, we can see the clear influence of the Roman Pantheon, and in its beautifully integrated piers, arches, and vaults, the legacy of Roman imperial-era structures.

- Centuries after the construction of Hagia Sophia, Byzantine architecture spread throughout the eastern Mediterranean and then back to the West in such important buildings as Charlemagne's imperial chapel at Aachen, Germany, and St. Mark's Basilica in Venice. In the modern era, these great structures have inspired the design of fine Neo-Byzantine buildings, such as the Cathedral Basilica of Saint Louis, completed in 1914, and the Los Angeles Public Library of 1926.

The Renaissance and Beyond

- Perhaps the most important source of technological continuity with the Classical world was the Italian **Renaissance** of the 14th century. This great revival of Classical literary and artistic values stimulated an intense interest in Greek and Roman architecture and engineering. Interest was greatly intensified in 1414, when a Florentine scholar named Poggio Bracciolini discovered a long-forgotten copy of Vitruvius's *De Architectura* in a medieval abbey in Switzerland.

- When Bracciolini shared this manuscript with the broader community of Renaissance scholars, artists, and builders, *De Architectura* quickly became the architectural and engineering bible of the Western world.

- In 1499, inspired by Vitruvius, a Dominican friar named Giovanni Giocondo reverse-engineered pozzolana-based concrete for the piers of his Notre Dame bridge in Paris—the first known use of concrete since the fall of Rome. Vitruvius also appears to have stimulated new interest in urban planning and substantially influenced the development of modern machinery.

- A drawing from an early-19th-century French text shows a construction crane that incorporates only a few minor improvements over one of Vitruvius's designs—and uses the same human-powered tread wheel as the larger Roman variant. Even Archimedes's screw remains in widespread use today. Its basic configuration is commonly used in wineries for moving grapes, in combine harvesters for moving grain, in modern snowblowers, and even in an electrical power plant in England.

- By the late 16th century, Renaissance architecture evolved toward a new form of expression: the **Baroque** style, which employed the Classical orders in a looser, more theatrical way. By the mid-18th century, however, the Baroque style had evolved into something so excessively ornate that it inevitably provoked a reaction.

Neoclassical Style

- This reaction came in the form of a new movement, called **Neoclassical**, which sought to restore the "pure" architectural forms of ancient Greece and Rome. Neoclassical architecture was also motivated by the mid-18th-century discoveries of Pompeii and Herculaneum, which sparked tremendous popular interest in ancient Roman civilization.

- Thus, when Thomas Jefferson designed his own home, Monticello, in 1772, he used the Roman Doric of Vitruvius, rather than the original Greek Doric of the Parthenon. Jefferson's stunning Virginia State Capitol of 1792 might appear to be Greek, but it was actually modeled on a well-preserved Roman temple in southern France. And Jefferson's Rotunda—the library at the University of Virginia—was clearly inspired by the Roman Pantheon.

- These buildings reflect the emergence of a uniquely American form of Neoclassical architecture called the **Federal style**, which combined Classical columns, pediments, and arches with simple red-brick walls to evoke the spirit of the Roman Republic. The Federal style is so common and so distinctly American that we easily overlook its direct connection with ancient Rome.

- In the late 18th century, the Ottomans gradually opened Greece to the West and stimulated a new phase of the Neoclassical called **Greek Revival**. Like the Federal style, Greek Revival architecture was not just limited to grand public buildings. Even today, it can be found in myriad private homes and small-town banks, libraries, and churches.

The Persistence of Classicism

- Neoclassical architecture remained popular through much of the 20th century, especially for buildings of particular cultural or political importance. The original Penn Station in New York City, completed in 1910, was faithfully modeled on the Baths of Caracalla—complete with Corinthian columns and coffered vaulting.

- Also consider the great monumental buildings of Washington, D.C.: the Lincoln Memorial of 1922, the National Gallery of 1937, the Jefferson Memorial of 1943. Who could deny the power of these Classical forms in connecting the American democratic republic with its distant roots in Athens and Rome?

- And while the Neoclassical style waned somewhat near the end of the 20th century, it appears to be staging a comeback in the 21st. The Schermerhorn Symphony Center in Nashville is a fine example of a 21st-century structure that has continued the Neoclassical tradition in bold new ways.

- Sackler Library at the University of Oxford is proof that the Classical character is more than just an aesthetic veneer. According to the Sackler Library's architect, the building meets today's stringent demands for sustainability and energy efficiency primarily through the use of locally available materials and traditional design features, such as thermally stable solid masonry walls and relatively small windows positioned for optimum light and ventilation. The concepts of using wall thickness, window size, and orientation to achieve energy efficiency come to us from Vitruvius, *De Architectura*, Book VI.

- Many ancient ideas still have a place in our modern technological systems—a fact that can only increase our admiration for the ancient engineers who conceived these innovations so many centuries ago.

Important Terms

aeolipile: The world's first steam engine, invented by Hero of Alexandria.

Baroque: An artistic and architectural style that followed the Renaissance and was characterized by exaggerated motion, drama, exuberance, and grandeur. Baroque architects drew on Classical forms but used them in a looser, more theatrical way than their Renaissance predecessors.

Byzantine Empire: The political entity that evolved from the eastern Roman Empire and continued through late antiquity and the Middle Ages. The capital of the Byzantine Empire was the city of Constantinople (modern Istanbul).

Federal style: An American adaptation of Neoclassical architecture, popular in the late 18th and early 19th centuries. The Federal style drew heavily on ancient Roman architectural forms.

Greek Revival: A late phase of the Neoclassical movement, stimulated when the Ottoman Turks gradually opened Greece to the West in the late 18th century.

Neoclassical: An architectural style of the 18th and 19th centuries. A reaction to the excesses of the Baroque era, Neoclassical architecture reflected a desire to return to the "pure" forms of ancient Greece and Rome.

Renaissance: The period from the 14th to the 16th century, when Europe experienced a great revival of interest in the literary and artistic values and forms of Classical antiquity.

Suggested Reading

Landels, *Engineering in the Ancient World*, chapter 1.

Nuttgens, *The Story of Architecture*.

Summerson, *The Classical Language of Architecture*.

Questions to Consider

1. What modern manifestations of Classical-era technology have you encountered in your daily life?

2. What ancient site would you most like to visit? How will you prepare yourself to make the most of this experience?

Glossary

actus: A Roman unit of land measurement, equal to 120 Roman feet.

aediculae: Covered niches, flanked by columns, intended as shelters for statues or shrines.

aeolipile: The world's first steam engine, invented by Hero of Alexandria.

agger: A ramped embankment used in Roman siege warfare.

aggregate: Sand, stone, or rubble used as a component of concrete.

agora: The central marketplace of a Greek city.

agrimensore: A Roman surveyor ("land measurer").

alloy: A metal composed of two or more elements.

amphora: A terra-cotta jar, used to transport and store liquids.

anathyrosis: Hollowing of the end faces of a stone block to facilitate a precise fit.

annular corridor: A ring-shaped corridor that runs around the circumference of a circular or elliptical structure.

antefix: A decorative boss used to hide the exposed ends of roof tiles along the lower edge of a roof.

apodyterium: A room for changing clothes in a Roman bath.

aquaclude: An impermeable stratum of soil or rock, over which an aquifer forms.

aqueduct: A manufactured structure that carries water from a distant source to a city or town.

aquifer: A geologic formation consisting of a porous stratum of soil or rock that is fully saturated with water.

arcade: An arrangement of multiple adjacent arches.

arch: A structural element that can span a horizontal distance while carrying load primarily in compression.

architekton: Greek term for the man who designed buildings and supervised their construction. (The equivalent Latin term is *architectus*.) The *architekton* served the modern functions of architect, structural engineer, and construction manager.

architrave: A rectangular beam spanning across the tops of two adjacent columns. The architrave is the lowest element of the entablature.

ashlar: A type of masonry construction consisting of rectangular stone blocks set in horizontal rows (or courses).

Baroque: An artistic and architectural style that followed the Renaissance and was characterized by exaggerated motion, drama, exuberance, and grandeur. Baroque architects drew on Classical forms but used them in a looser, more theatrical way than their Renaissance predecessors.

barrel vault: A vault with the shape of a half-cylinder.

basilica: A Roman public building characterized by a large covered central hall.

batten: A lightweight beam that directly supports a row of roof tiles.

beam: A structural element that carries load primarily in bending.

bearing: In surveying, a line oriented in a particular direction and passing through a designated reference point.

bessalis: An 8-inch-square Roman brick.

bipedalis: A 2-foot-square Roman brick.

bireme: A warship with oars on two levels.

block-and-tackle system: A system of pulleys and ropes, which provides mechanical advantage for lifting.

bollard: A vertical post attached firmly to the ground.

braces: A pair of ropes extending from the tip of a yardarm to the ship's stern; used to rotate the yardarm horizontally.

brails: Light ropes attached to the lower edge of the sail that extend upward over the yardarm and then back to the stern, where they are tied or held by a crewman; used to adjust the size and shape of the sail.

bridge tree: A horizontal beam that supports the vertical shaft of a mill and allows adjustment of the upper millstone's height.

bronze: An alloy of copper and tin.

bucket chain: A human-powered water-lifting device consisting of a string of buckets fixed to a pair of iron chains, driven by a rotating shaft.

bucket wheel: A human-powered water-lifting device consisting of a rotating wheel with buckets positioned around its outer rim.

buckling: Stability failure of a structural element subjected to compression.

buoyancy: The scientific principle governing floating objects. When an object floats, the weight of the water it displaces is equal to the weight of the object.

buttress: A pier or thickened section of a wall that resists the lateral thrust of a vault or dome.

buttress vault: A barrel vault used to restrain the lateral thrust of a larger vault, arch, or dome.

Byzantine Empire: The political entity that evolved from the eastern Roman Empire and continued through late antiquity and the Middle Ages. The capital of the Byzantine Empire was the city of Constantinople (modern Istanbul).

calcium carbonate: A solid white substance that accumulates on the inside surfaces of pipes carrying water with high mineral content.

caldarium: The hot room in a Roman bath.

calix: A standardized bronze fitting used to regulate the amount of water supplied to individual users in a Roman water distribution system.

capital: The decorative top of a column.

capstan: A device for harnessing and amplifying human or animal power, to apply tension to a rope. A capstan has a vertical shaft, while a windlass typically has a horizontal shaft.

cardo maximus: The main north-south street in a Roman town or city.

carruca: A large four-wheeled Roman traveling coach.

castellum divisorium: A structure that divides the incoming flow from an aqueduct into multiple channels for subsequent distribution.

casting: The process of forming a metal into a desired shape by heating it to the melting point and then pouring the molten material into a mold.

castrum: A Roman military camp.

catillus: The hourglass-shaped upper stone in a rotary mill; the *catillus* rotates on the lower stone, or *meta*.

cavea: The seating area of a Roman amphitheater.

cella: Enclosed sanctuary within a Greek temple. The *cella* housed the cult statue of the god to whom the temple was dedicated.

centuria: In Roman surveying, a square plot of land measuring 20 *actus* by 20 *actus*.

centuriation: The Roman practice of subdividing conquered territory into regular square parcels.

charcoal: A fuel produced by heating hardwood in a reduced-oxygen environment to drive off water and resins. Because charcoal is nearly pure carbon, it is capable of burning at much higher temperatures than wood.

cheiroballistra: A small, mobile Roman arrow-shooting catapult, which used iron rather than wood for its main frame.

chorobates: A Roman surveying device used to measure the change in elevation over a distance.

cisium: A Roman two-wheeled carriage, used as a taxi.

cistern: A masonry tank, usually located just below ground level and used to collect rainwater from a roof or paved surface.

coffer: A polygonal indentation in a vault, dome, or ceiling.

cofferdam: A temporary structure used to construct bridge piers and port facilities underwater.

collier: A person who produces charcoal.

colonnade: A row of columns.

column: A structural element that carries load primarily in compression.

composite bow: A bow constructed from three different materials—animal sinew (in tension) and bone (in compression), around a wooden core.

compound machine: A system composed of multiple simple machines.

compression: An internal force or stress that causes shortening of a structural element.

concrete: A manufactured structural material created by combining cement, water, and aggregate.

Corinthian order: An architectural style characterized by relatively slender columns and ornate capitals decorated with stylized acanthus leaves.

cornice: The uppermost element of the entablature in a Greek temple. The cornice projects outward to protect the structure from the elements.

course: A horizontal row of cut stones or bricks.

crepidoma: The three- or four-step stone platform on which a Greek temple was built.

cross-section: The geometric shape of a structural element, viewed from its end.

cursus publicus: The official Roman state courier system, established by Emperor Augustus.

cut and cover: Method used to construct an underground aqueduct channel just below ground level.

cyclopean: A type of masonry construction consisting of very large polygonal stones fitted closely together without mortar.

dactyl: A Greek unit of measurement, equal to approximately ¾ inch.

decempeda: A Roman measuring rod, made of hardwood and capped with bronze fittings that allowed multiple rods to be connected end to end.

decumanus maximus: The main east-west street in a Roman town or city.

dentatum ("toothed disk"): Roman term for a gear.

diffusion: The process by which an innovation is communicated through a social system.

Doric order: An architectural style characterized by relatively stout columns and simple, unadorned capitals.

dowel: A cylindrical or rectangular iron peg inserted into the top and bottom surfaces of stone blocks to align and connect them.

drum: A cylindrical segment of a stone column or a cylindrical wall that supports a dome.

durability: The capacity of a material to resist deterioration by weathering, corrosion, or rot.

elastic energy: Energy stored in the deformation of a material—e.g., the bending of a bow (also called strain energy).

ellipse: The geometric shape formed by a set of points that are the same total distance from two points, called foci.

empirical design: The process of designing a structure, machine, or system by making a succession of well-reasoned incremental changes based on experience and the observed performance of previous designs.

engaged column: A decorative half-column projecting from a wall or pier.

engineering: The application of math, science, and technology to create a structure, device, machine, system, or process that meets a human need.

entablature: The architectural element spanning across the tops of columns in Greek architecture. The entablature consists of three parts—the architrave, frieze, and cornice.

entasis: Slight bulge in the shape of a Greek column, incorporated to enhance the column's appearance.

euthytone: An arrow-shooting torsion catapult.

Federal style: An American adaptation of Neoclassical architecture, popular in the late 18th and early 19th centuries. The Federal style drew heavily on ancient Roman architectural forms.

felloe: The outer rim of a spoked wheel.

flat arch: A flat, horizontal structural element that carries load as an arch; also called a lintel arch.

flow rate: The output of a water-lifting device, expressed in volume per unit of time (e.g., gallons per hour).

fluted: Characterized by vertical grooves carved into the outer surface of a stone column.

flying buttress: A structural element that resists the lateral thrust of ceiling vaults. Flying buttresses are external to the structure they support.

force: A push or pull applied to an object. A force is defined in terms of both magnitude and direction.

force pump: A human-powered water-lifting device that uses one or more pistons, cylinders, and valves to pump water through reciprocating (up-and-down) motion.

force-draw curve: A graph of the force required to draw a bow versus the distance drawn.

forum: The central public marketplace in a Roman town or city.

frieze: A decorative horizontal band forming one element of the entablature in a Greek temple.

frigidarium: The cold room in a Roman bath.

gastraphetes ("belly bow"): An arrow-shooting forerunner of the catapult. The *gastraphetes* incorporated a stiff bow and a mechanical configuration that allowed the archer to cock the weapon by leaning his full weight on the stock.

gradient: The slope of a road or aqueduct channel, expressed as a percentage: elevation change per horizontal distance.

Greek Archaic period: Historical period that began around 800 B.C. and ended with the emergence of Athenian democracy around 500 B.C.

Greek Revival: A late phase of the Neoclassical movement, stimulated when the Ottoman Turks gradually opened Greece to the West in the late 18th century.

groin vault: A vault formed by the intersection of two perpendicular barrel vaults.

groma: A Roman surveying instrument, used for laying out a rectangular grid.

groundwater: Water that infiltrates deeply into the soil and is ultimately collected in an aquifer.

gunwale: The heavy wooden rail at the top edge of a ship's hull.

halyard: A pair of ropes from which the yardarm is suspended.

hamaxa: A small two-wheeled, all-purpose carriage used by the Greeks.

header tank: The water reservoir at the start of an inverted siphon.

headrace: An artificial channel that delivers water to a waterwheel.

helepolis ("city taker"): A mobile siege tower.

helix: The spiral shape associated with a screw.

Hellenic period (also called Classical Greece): Historical period that began with the emergence of Athenian democracy around 500 B.C. and ended with the death of Alexander the Great in 323 B.C.

Hellenistic period: Historical period that began with the death of Alexander the Great in 323 B.C. and ended with the Roman conquest of Egypt in 30 B.C.

hematite: A type of iron ore.

hemicycle: A semicircular architectural feature.

heredia: In Roman surveying, a square plot of land measuring 2 *actus* by 2 *actus*.

hexastyle: A type of Greek temple with six columns across its front colonnade.

hodometer: A cart-mounted device used to measure long distances over land.

hogging: The bending of a ship's hull caused by a large wave amidships. A hogging hull bends concave downward.

hoplite: A heavily armored Greek infantryman, equipped for fighting in the phalanx formation.

hopper mill: A simple milling device that uses a lever to move the upper stone across the lower one.

hydraulic gradient: An imaginary line connecting the water surfaces of the header and receiving tanks in an inverted siphon system; used to calculate pressure in the pipeline.

hydrologic cycle: The natural process by which water falls to the earth in the form of precipitation; flows over and into the soil; is transported by streams and rivers; is stored in lakes and oceans; and ultimately, returns to the air by evaporation and transpiration.

hypocaust: The system used to heat both water and air in a Roman bath.

hypogeum: A two-story network of corridors, rooms, ramps, and shafts located underneath the wooden arena floor of the Colosseum.

hypozoma: A loop of heavy rope extending from the bow to the stern of a Greek trireme. When tightened by twisting, the ship's two *hypozomata* strengthened the hull against structural failure by hogging.

impost block: A stone block that supports the base of an arch.

inclined plane: A simple machine that allows an object to be lifted with an applied force less than the object's weight.

infrastructure: Large-scale technological systems that enhance societal functions, facilitate economic development, and enhance quality of life.

innovation: The process by which an invention is brought into use.

invention: The act of implementing an original idea in a new device.

inverted siphon: A type of aqueduct (or segment of an aqueduct) used to transport water across a valley through a pipeline under pressure.

Ionic order: An architectural style characterized by relatively slender columns and double-scroll capitals.

jugerum: In Roman surveying, a rectangular plot defined as the area of land that could be plowed by one pair of oxen in one day.

keel: The structural backbone of a ship.

kinetic energy: The energy associated with a mass in motion.

lever: A simple machine that magnifies an applied force.

Lewis bolt: A wedge-shaped apparatus used to lift stone blocks.

lifts: A pair of ropes extending from the tip of a yardarm to the top of the mast; used to support the yardarm and tilt it vertically.

lime mortar: A type of mortar manufactured by baking limestone in a kiln.

limestone: A sedimentary rock, usually formed from accumulated skeletal fragments of marine organisms, such as coral.

lintel: A horizontal structural element that spans an opening, such as a door or window.

locking scarf joint: An elaborate joint used to splice two pieces of wood together for a ship's keel.

longship: A ship with a long, narrow hull, powered primarily by rowers and generally used as a warship or pirate vessel.

machine: An assembly of fixed or moving parts, used to perform work.

malleability: The extent to which a material can be shaped or formed into thin sheets by hammering. Metals are generally very malleable, in contrast with stone, clay, and concrete.

marble: A metamorphic rock, created when limestone is subjected to intense heat and pressure deep below the earth's surface.

mechanical advantage: The amplification of an applied force by a mechanical device.

mechanical properties: Characteristics of a material that describe how the material responds to forces.

meta: The cone-shaped lower stone in a rotary mill; the upper stone, or *catillus*, rotates on the *meta*.

metope: A decorative element in the frieze of a Doric temple. Metopes are always alternated with triglyphs in the Doric frieze.

mina: A Greek unit of weight, equal to approximately 1 pound.

mortar: A substance used to fill the gaps between stones or bricks in masonry construction.

mortise-and-tenon joint: A joint used to fasten planks to each other in a shell-first hull.

Museum of Alexandria: A Hellenistic institution of learning, research, and invention, established by the Ptolemaic kings of Egypt around 300 B.C. The word *museum* refers to a "house of the Muses," rather than a museum in the modern sense.

mutule: A decorative element that represents the ends of angled roof rafters in a Greek temple.

natatio: A Roman open-air swimming pool.

Neoclassical: An architectural style of the 18th and 19th centuries. A reaction to the excesses of the Baroque era, Neoclassical architecture reflected a desire to return to the "pure" forms of ancient Greece and Rome.

noria: A water-powered bucket wheel.

octastyle: A type of Greek temple with eight columns across its front colonnade.

oculus: The circular opening at the top of a dome.

onager: A late-Roman catapult with only one throwing arm.

open-channel aqueduct: An aqueduct in which the water flows on a continuous downhill gradient and does not flow under pressure.

opus caementicium: Roman term for solid concrete.

opus incertum: A Roman wall construction system, consisting of outer facings of random fitted stones surrounding a concrete core.

opus quadratum: Roman term for ashlar stone construction.

opus recticulatum: A Roman wall construction system, consisting of outer facings of pyramid-shaped stones surrounding a concrete core.

opus testaceum: A Roman wall construction system, consisting of outer facings of overlapping triangular bricks surrounding a concrete core.

overshot wheel: A waterwheel that is mounted on a horizontal shaft and driven by water flowing into buckets at the top of the wheel.

oxidation-reduction: A chemical reaction in which one substance loses electrons (and is said to be oxidized) and another gains electrons (and is said to be reduced).

palaestra: A Roman open-air exercise area.

palintone: A stone-shooting torsion catapult (also called a ballista by the Romans).

pediment: Triangular gable on the front and rear of a Greek temple.

penteconter: A 50-oared longship.

periteichismos: A Greek siege tactic, in which the besieging force surrounded and isolated the besieged city with a substantial wall of wood or stone.

phalanx: A battlefield formation consisting of multiple ranks of closely spaced, heavily armored infantry called hoplites.

plank-on-frame: Construction method used in most wooden ships after the Classical era. The plank-on-frame hull was created by assembling the keel, stem post, sternpost, and wooden frames into a rigid framework. Planking was added only after this framework was complete. Planks were attached to the frames but not to each other.

plumbarius: A Roman plumber.

point: An angle equal to 1/32 of a full circle (or 11¼ degrees).

polis: The Greek city-state.

polycentric oval: The geometric shape formed by a series of interconnected circular arcs.

Pompeiian mill: An animal-powered rotary mill (also called a donkey mill).

portico: The covered front porch of a Classical-era building.

power: The rate at which work is done.

pozzolana: A naturally occurring volcanic ash used as the cement in Roman concrete.

praefurnium: The wood-burning furnace that supplied heat in a hypocaust.

principia: The headquarters building at the center of a Roman *castrum*.

prop: A vertical strut forming one element of the prop-and-lintel roof system.

prop-and-lintel: The timber roof system used in most Greek temples (and other contemporary structures).

pulley: A simple machine that changes the direction of a force.

purlin: A longitudinal beam that supports the rafters in a prop-and-lintel roof system.

qanat: A tunnel driven into a hillside to tap an underground aquifer.

quoin: A rectangular stone used to reinforce the corner of a building.

rafter: An angled beam that supports a roof.

raking vault: A barrel vault inclined at an angle.

receiving tank: The water reservoir at the end of an inverted siphon.

relieving arch: An arch built into a wall above a door or window opening to divert compressive force around the opening.

Renaissance: The period from the 14th to the 16th century, when Europe experienced a great revival of interest in the literary and artistic values and forms of classical antiquity.

ridge beam: A longitudinal beam that supports the peak of a roof.

right-angle gearing: An assembly of two gears that converts horizontal shaft power to vertical shaft power in a mill.

Roman Empire, imperial Rome: Historical period that is generally considered to have begun with Octavian's assumption of the title *Augustus* in 27 B.C. and ended in A.D. 476, when the last western Roman emperor was deposed.

Roman Republic: Historical period that began with the establishment of the Roman Republic in 509 B.C. and is generally considered to have ended with Octavian's assumption of the title *Augustus* in 27 B.C.

rotary quern: A milling device consisting of a dome-shaped lower stone, a concave upper stone with a hopper carved into its center, and a wood or iron pivot (called the spindle) on which the upper stone rotated.

roundship: A ship with a broad, rounded hull, powered primarily by sail and generally used for carrying cargo.

saddle quern: A simple milling device consisting of two shaped stones.

sagging: The bending of a ship's hull caused by large waves fore and aft and a trough amidships. A sagging hull bends concave upward.

saqiya: An animal-powered tympanum.

scorpion: A Roman arrow-shooting catapult with forward-curving arms.

screw pump: A human-powered water-lifting device consisting of one or more helical vanes rotating on an inclined shaft.

screw: A simple machine that converts a rotational force (or torque) to a linear force.

secondary *castellum*: A structure that distributes an incoming flow of water to multiple users while also controlling pressure in the water distribution system.

semi-dome: A half-dome.

settling tank: A masonry tank used to remove sand, silt, and other suspended solids from aqueduct-supplied water before it entered the urban distribution network.

shadoof: A primitive water-lifting device, consisting of a counterweighted beam pivoting on a vertical post.

sheet: A pair of ropes extending from the lower corners of the sail to the helmsman's position at the stern of the ship; used to control the sail.

shell-first: Construction method used for most ancient ship hulls. The shell-first hull was created by attaching individual planks to each other along their edges, using mortise-and-tenon joints. Light wooden frames were added to the inside of the hull only after it was completely formed.

shrouds: Ropes that support a ship's mast in the port and starboard directions.

siege: A form of warfare aimed at overcoming fortified defensive positions. Siege warfare often employed heavy machinery, such as mobile towers, battering rams, and catapults.

smelting: The process of heating an ore to produce a usable metal.

soil water: Water that infiltrates into the earth but is retained near the surface, in the voids between soil particles.

solder: A mixture of lead and tin used to connect lead pipes.

spindle: The iron pivot on which a millstone rotates.

stays: Ropes that support a ship's mast in the fore and aft directions.

step: The socket in a ship's keel, used to support the mast.

stoa: A long, narrow Greek building used to house shops and offices, usually located in the agora.

strength: The maximum stress a material can withstand before it breaks. Strength can be defined for both tension and compression.

stress: The intensity of internal force within a structural element, defined in terms of force per area (pounds per square inch).

structure: A technological system—typically a building, bridge, or tower—that is designed to carry load.

stylobate: The top level of the *crepidoma*.

substructio: A stone- or brick-faced embankment used to support an aqueduct channel a few feet above ground level.

surface water: Water that flows over the earth's surface and in rivers and streams.

tacking: A sailing maneuver in which the ship turns its bow through the wind in order to receive the wind on its opposite side. By executing this maneuver repeatedly, the vessel can make progress directly into the wind, albeit along a zigzag path.

tailrace: An artificial channel that carries water away from a water wheel.

technology in use: The processes of employing and maintaining existing technologies, and adapting them to new purposes over time.

technology: A human-made structure, device, machine, system, or process that meets a human need. Technology is the product of engineering.

tension: An internal force or stress that causes elongation of a structural element.

tepidarium: The medium-temperature room in a Roman bath.

terra-cotta: A ceramic material created by firing clay in a kiln to improve its strength and durability.

terrain profile: A graph showing the variation in elevation along a route.

testudo ("tortoise"): A Roman infantry formation used for assaulting fortified positions. Soldiers formed a testudo by locking their shields together over their heads to form a protective barrier.

thrust: The outward force generated by an arch under load; also, the tendency of an arch to spread out laterally under load.

tie-beam truss: A structural system composed of members configured in interconnected triangles.

torque: The tendency of a force to cause rotation, expressed in terms of force times distance.

torsion: Twisting of a structural or mechanical element.

torsion catapult: A weapon that used two vertically oriented torsion springs and two throwing arms to shoot heavy arrows (called bolts) or stones.

torsion spring: A spring consisting of a bundle of rope made from animal sinew. Elastic energy is stored in the spring by twisting it.

tortoise: An armored wheeled vehicle used in siege warfare.

trabeated: A type of structural system consisting of beams supported on columns.

travertine: A strong, hard limestone commonly used in Roman construction.

tread wheel: A device for harnessing and amplifying human power in construction cranes, water-lifting devices, and similar machines.

trestle: A wooden frame that supports the deck of a timber bridge structure.

triglyph: A decorative element in the frieze of a Doric temple. Triglyphs are always alternated with metopes in the Doric frieze.

trireme: In antiquity, a warship with oars on three levels. In later eras, other trireme configurations were also used, for example, one level of oars with three rowers per oar.

tubuli: Terra-cotta pipes used to heat the walls in a hypocaust system.

tufa: A soft, porous limestone commonly used in Roman construction.

tympanon: Decorative triangular panel within the pediment of a Greek temple.

tympanum: A human-powered water-lifting device consisting of a compartmented drum rotating on a horizontal shaft.

undershot wheel: A waterwheel that is mounted on a horizontal shaft and driven by water passing underneath the wheel.

urban planning: The science of designing the overall layout, functional organization, and architectural character of a city or town.

vault: An arched roof or ceiling.

velarium: The rope-and-canvas awning used to shade spectators at Roman amphitheaters.

vertical-shaft wheel: A waterwheel that is mounted on a vertical shaft and is driven by water striking the wheel's angled vanes at high velocity.

vinea: Hide-covered wooden frames linked together to create a protected corridor, through which workmen and assault troops could approach a fortified position in comparative safety.

voussoir: A wedge-shaped component of an arch.

water table: The upper surface of an aquifer.

weather helm: The tendency of a ship sailing across the wind to turn into the wind.

wedge: A simple machine that converts a single force into a pair of opposing forces that are oriented perpendicular to the surfaces of the wedge.

wheel and axle: A simple machine that facilitates horizontal movement with a minimum application of force.

windlass: A device for harnessing and amplifying human power, to apply tension to a rope. A windlass typically has a horizontal shaft, while a capstan has a vertical shaft.

work: The quantity of energy expended when a force moves through a distance.

workability: The extent to which a material can be molded, carved, or deformed to attain a desired physical shape.

wrought iron: Iron produced by hammering the product of the smelting process (called a bloom) to drive out impurities.

yardarm: A horizontal spar from which a sail is suspended.

yoke: A heavy wooden beam fitted over the shoulders of a draft animal and held in place with a harness.

Bibliography

Comprehensive References on Ancient Technology

De Camp, L. Sprague. *The Ancient Engineers*. New York: Ballantine, 1963.

Hodges, Henry. *Technology in the Ancient World*. New York: Barnes and Noble, 1970.

Humphrey, John W., John P. Oleson, and Andrew N. Sherwood. *Greek and Roman Technology: A Source Book*. London: Routledge, 1998.

Landels, J. G. *Engineering in the Ancient World*. New York: Barnes & Noble, 1978.

Oleson, John Peter, ed. *The Oxford Handbook of Engineering and Technology in the Classical World*. New York: Oxford University Press, 2008.

Scarre, Chris. *Timelines of the Ancient World: A Visual Chronology from the Origins of Life to AD 1500*. New York: Dorling Kindersley, 1993.

White, K. D. *Greek and Roman Technology*. Ithaca: Cornell University Press, 1984.

Primary Sources

Caesar, Julius. *The Gallic War*. Translated by Carolyn Hammond. New York: Oxford University Press, 1996.

Josephus. *The Jewish War*. Translated by G. A. Williamson. New York: Penguin, 1981.

Palladio, Andrea. *The Four Books of Architecture*. Translated by Isaac Ware. Mineola, NY: Dover, 1965.

Vitruvius. *The Ten Books on Architecture*. Translated by Morris Hicky Morgan. New York: Dover, 1960.

Principles of Mechanics and Structural Engineering
Gordon, J. E. *Structures: Or Why Things Don't Fall Down*. New York: Da Capo, 1978.

Hilson, Barry. *Basic Structural Behavior: Understanding Structures from Models*. London: Thomas Telford, 1993.

Riley, William F., Leroy D. Sturges, and Don H. Morris. *Statics and Mechanics of Materials: An Integrated Approach.* New York: John Wiley and Sons, 1995.

Salvadori, Mario. *Why Buildings Stand Up: The Strength of Architecture*. New York: Norton, 1980.

———, and Matthys Levy. *Why Buildings Fall Down*. New York: Norton, 1992.

Materials
Liebson, Milt. *Direct Stone Sculpture*. Atglen, PA: Schiffer, 2001.

Sass, Stephen L. *The Substance of Civilization: Materials and Human History from the Stone Age to the Age of Silicon*. New York: Arcade, 1998.

Spencer, Edward B. "Stones Used in the Construction and Decoration of Ancient Rome." *The Classical Journal*, no. 33.5 (1938): 271–279.

General References on Architecture and Construction
Addis, Bill. *Building: 3000 Years of Design, Engineering and Construction*. London: Phaidon, 2007.

Burden, Ernest. *Illustrated Dictionary of Architecture*. New York: McGraw Hill, 2002.

Norwich, John J. *Great Architecture of the World*. London: Mitchell Beazley, 1975.

Nuttgens, Patrick. *The Story of Architecture*. Oxford: Phaidon, 1983.

Summerson, John. *The Classical Language of Architecture*. London: Thames and Hudson, 1983

Greek and Roman Architecture and Construction

Adam, Jean-Pierre. *Roman Building: Materials and Techniques*. Indianapolis: Indiana University Press, 1994.

American School of Classical Studies at Athens. *Ancient Athenian Building Methods*. Prepared by John M. Camp and William B. Dinsmoor. Lunenburg, Vermont: The Stinehour Press, 1984.

Coulton, J. J. *Ancient Greek Architects at Work*. Ithaca: Cornell University Press, 1977.

———. "Lifting in Early Greek Architecture." *The Journal of Hellenic Studies*, no. 94 (1974): 1–19.

DeLaine, Janet. "Structural Experimentation: The Lintel Arch, Corbel, and Tie in Western Roman Architecture." *World Archaeology*, no. 21.3 (1990): 407–424.

Hodge, Trevor. *The Woodwork of Greek Roofs*. Cambridge: Cambridge University Press, 1960.

Krautheimer, Richard. "The Constantinian Basilica." *The Dumbarton Oaks Papers*, no. 21 (1967): 115–140.

Lancaster, Lynne. "Building Trajan's Markets." *American Journal of Archeology*, no. 102.2 (1998): 283–308.

Lawrence, Arnold W. and Richard Allan Tomlinson. *Greek Architecture*. New Haven: Yale University Press, 1996.

MacDonald, William L. *The Architecture of the Roman Empire*. New Haven: Yale University Press, 1965.

———. *The Pantheon: Design, Meaning, and Progeny*. Cambridge, MA: Harvard University Press, 1976.

Malacrino, Carmelo G. *Constructing the Ancient World: Architectural Techniques of the Greeks and Romans*. Los Angeles: Getty Publishing, 2010.

Mark, Robert, and Paul Hutchinson. "On the Structure of the Roman Pantheon." *The Art Bulletin*, no. 68.1 (1986): 24–34.

Moore, David. "The Pantheon." http://www.romanconcrete.com/docs/chapt01/chapt01.htm.

Muller, Valentin. "The Roman Basilica." *American Journal of Archaeology*, no. 41.2 (1937): 250–261.

Spawforth, Tony. *The Complete Greek Temples*. London: Thames & Hudson, 2006.

Taylor, Rabun. *Roman Builders: A Study in Architectural Process*. Cambridge: Cambridge University Press, 2003.

Ulrich, Roger B. *Roman Woodworking*. New Haven, CT: Yale University Press, 2007.

Infrastructure and Urban Planning

Cavaglieri, Giorgio. "Outline for a History of City Planning from Prehistory to the Fall of the Roman Empire." *Journal of the Society of Architectural Historians*, no. 8.1 (1949): 43–54.

Chanson, H. "Hydraulics of Roman Aqueducts: Steep Chutes, Cascades, and Dropshafts." *American Journal of Archaeology*, no. 104 (2000): 47–72.

Davies, Hugh. "Designing Roman Roads." *Brittannia*, no. 29 (1998): 1–16.

DeLaine, Janet. *The Baths of Caracalla: A Study in the Design, Construction, and Economics of Large-Scale Building Projects in Imperial Rome*. Portsmouth, RI: Journal of Roman Archaeology, 1997.

Fagan, Garrett G. "The Genesis of the Roman Public Bath: Recent Approaches and Future Directions." *American Journal of Archeology*, no. 105.3 (2001): 403–426.

Haverfield, F. *Ancient Town Planning*. New York: Oxford University Press, 1913.

Hodge, Trevor A. *Roman Aqueducts and Water Supply*. London: Duckworth, 2002.

Macauley, David. *City: A Story of Roman Planning and Construction.* Boston: Houghton Mifflin, 1974.

Piranomonte, Marina. *The Baths of Caracalla*. Milan: Electa, 1998.

Staccioli, Romolo. *The Roads of the Romans.* Los Angeles: Getty Trust Publications, 2004.

Ward-Perkins, J. B. "Early Roman Towns in Italy." *The Town Planning Review*, no. 26.3 (1955): 126–154.

Machines

Lancaster, Lynne. "Building Trajan's Column." *American Journal of Archeology*, no. 103.3 (1999): 419–439.

Oleson, John P. *Greek and Roman Mechanical Water-Lifting Devices: The History of a Technology*. Buffalo: University of Toronto Press, 1984.

Weller, Judith A. "Roman Traction Systems." http://www.humanist.de/rome/rts/index.html

Wilson, Andrew. "Machines, Power and the Ancient Economy." *The Journal of Roman Studies*, no. 92 (2002): 1–32.

Military Technology

Baatz, Dietwulf. "Recent Finds of Ancient Artillery." *Britannia*, no. 9 (1978): 1–17.

Campbell, Duncan. *Besieged: Siege Warfare in the Ancient World.* New York: Osprey, 2006.

———. *Greek and Roman Siege Machinery, 399 BC–AD 363*. New York: Osprey, 2003.

———. *Greek and Roman Artillery, 399 BC–AD 363*. New York: Osprey, 2003.

Marsden, E. W. *Greek and Roman Artillery: Historical Development*. Oxford: Oxford University Press, 1969.

———. *Greek and Roman Artillery: Technical Treatises*. Oxford: Oxford University Press, 1971.

Ships and Seafaring

Casson, Lionel. *Ships and Seamanship in the Ancient World.* Baltimore: The Johns Hopkins University Press, 1971.

Hale, John R. *Lords of the Sea*. New York: Viking, 2009.

Harland, John. *Seamanship in the Age of Sail*. London: Conway Maritime Press, 1984.

Morrison, J. S., J. F. Coates, and N. B. Rankov. *The Athenian Trireme*. New York: Cambridge University Press, 2000.

Strauss, Barry. *Salamis: The Naval Encounter That Saved Greece—and Western Civilization*. New York: Simon and Schuster, 2004.